Bioreactor Implementation in the Agro-Food Industries

SCIENCES

Agronomy and Food Science,
Field Directors – Jack Legrand and Gilles Trystram

Food Bioprocesses, Subject Head – Mohamed Ghoul

Bioreactor Implementation in the Agro-Food Industries

Technology, Kinetics and Modelization

Coordinated by
Mohamed Ghoul

iSTE

WILEY

First published 2024 in Great Britain and the United States by ISTE Ltd and John Wiley & Sons, Inc.

ISTE Ltd
27-37 St George's Road
London SW19 4EU
UK

www.iste.co.uk

John Wiley & Sons, Inc.
111 River Street
Hoboken, NJ 07030
USA

www.wiley.com

Library of Congress Control Number: 2024943819

British Library Cataloguing-in-Publication Data
A CIP record for this book is available from the British Library
ISBN 978-1-78945-155-9

ERC code:
LS9 Biotechnology and Biosystems Engineering
LS9_5 Food biotechnology and bioengineering
SH1 Individuals, Markets and Organisations
SH1_12 Environmental economics; resource and energy economics; agricultural economics

Contents

Preface

Mohamed GHOUL
Laboratoire Réaction et Génie des Procédés (LRGP),
CNRS, Université de Lorraine, Nancy, France

The use of bioreactors for the production of food ingredients has been burgeoning in recent years. These processes enable the creation of a controlled environment adapted to the specific needs of each microorganism, while minimizing the environmental impact of these processes. However, the implementation of bioreactors on an industrial scale often remains empirical based on the experience and expertise of individual companies.

A rational approach to optimal bioreactor implementation is possible, but requires a mastery of a number of scientific concepts relating to material and heat balances, microbial kinetics and their implementation. Indeed, depending on the kinetics of the microorganism used, the composition of the reaction medium and the nature of the ingredients sought, several configurations and modes of bioreactor operation are possible.

The aim of this book is to provide an exhaustive and accurate presentation of all of these concepts, making them accessible to students, researchers and professionals wishing to exploit the potential of bioreactors in order to produce food ingredients and develop innovative processes.

The first part of this book therefore develops the fundamentals of bioreactor engineering, including material and heat balances, microbial kinetics, bioreactor implementation, as well as agitation and mixing techniques. This knowledge is essential for controlling the biological and physicochemical processes that take place in bioreactors. These processes can have contradictory effects, and the search for compromises is often necessary to achieve optimal operation of these processes. Criteria for optimizing bioreactor operation (batch, fed-batch or continuous) and choosing the most suitable agitation technology for the biological system used (bacteria, yeast, fungi, microalgae), as well as the rheology of the culture medium, will be presented and discussed.

The second part illustrates the use of bioreactors for the production of several ingredients and metabolites of interest. In this way, we discover how to integrate all of the knowledge developed in the first part of this book in an original and innovative way, so as to produce (i) biomass and food bioactives by cultivating different species of microalgae, (ii) glutamic acid by exploiting the potential of bacteria in this field, (iii) beverages such as beer thanks to the fermentative power of yeast and (iv) ethanol from renewable bioresources by yeast fermentation.

This book is aimed at professionals, researchers and students alike, and is therefore a basic reference in the field of bioreactors applied to food bioprocesses.

Mohamed GHOUL
October 2024

Introduction

Jean-Marc ENGASSER and Mohamed GHOUL
Laboratoire Réaction et Génie des Procédés (LRGP), CNRS, Université de Lorraine, Nancy, France

Processes for the production of biomass and metabolites by different species of microorganisms are key techniques in many industries, including food processing. Microorganisms such as bacteria, yeasts, fungi and microalgae are widely used in these processes. They act as "cellular factories", performing numerous enzymatic reactions to transform various substrates into the desired products.

In most cases, specific strains of microorganisms are selected for their production performance. These strains are often genetically modified to improve their ability to produce specific metabolites. Indeed, genetic engineering and metabolic engineering techniques make it possible to create new industrial microorganisms that are capable of synthesizing specific metabolites. These new strains can significantly increase productivity levels. In some cases, mixed cultures are used to optimize processing activities and yields. This approach can be particularly useful when different strains have complementary roles.

Processes using microorganisms are used in the food industry for a variety of applications, including the production of proteins, foods, beverages, colorants, acids and other food-related bioactives.

The products resulting from these transformations fall into four distinct categories:

– *small molecules* resulting from cellular metabolism, such as organic acids, amino acids, nucleotides, alcohols, esters, dyes and vitamins;

– *macromolecules* such as polysaccharides, peptides and proteins;

– *whole cells* such as baker's yeast, microalgae, lactic ferments and probiotics;

– *fermented food products* such as beer and yogurt.

In addition to their use in food production, these processes also have numerous environmental applications in waste decontamination. Common examples include activated sludge processes, denitrification and anaerobic digestion for methane production. These processes can also contribute to environmental sustainability, as they can use renewable substrates and produce less waste than other chemical methods.

Bioreactors are essential equipment in these processes in order to enable precise control of microorganism growth conditions such as temperature, pH, agitation and nutrient concentration.

In addition to the central operation involving the bioreactor, biotransformation processes often involve a number of upstream and downstream unit operations to efficiently transform raw materials into high-quality end products.

Upstream, raw materials may require pretreatment. This may include grinding, cutting, washing or other steps to prepare the raw materials. To avoid contamination by undesirable microorganisms, the culture media used in the bioreactor must be sterilized. This can be achieved by autoclaving, sterilizing filtration or through other sterilization methods.

In the downstream step, after bioconversion, the final products are often mixed with the culture medium and microbial cells. The first step usually involves separation operations to isolate the desired products. This may include centrifugation, filtration, decantation or other separation methods, depending on the characteristics of the products and the medium. Isolated products may contain impurities such as undesirable compounds, residual cells or contaminants. Purification aims to remove these impurities to obtain a pure and high-quality product. Common purification techniques include chromatography, precipitation, ion exchange and other product-specific methods. Once products are purified, they may require formulation to achieve the desired final composition. This may include steps such as adding stabilizers, preservatives, colorants or other additives to meet final product specifications.

These upstream and downstream operations are essential to guarantee the quality, purity and conformity of the final products to food industry standards. The exact sequence of operations may vary according to product, process and specific application requirements, but they are all crucial to the overall success of the biotransformation process.

For optimal implementation of processes involving bioreactors with different species of microorganisms to produce metabolites of interest, it is therefore necessary to not only master the biotransformation stage, but also the upstream and downstream stages of the process. This calls for a multidisciplinary approach, with a mastery of a number of engineering and life science concepts. All of these bioreactor engineering tools will be presented and illustrated by practical examples from the food industry.

PART 1

The Fundamentals of Fermentation and Photobioreactor Engineering

Introduction to Part 1

Jean-Marc ENGASSER and Mohamed GHOUL
Laboratoire Réaction et Génie des Procédés (LRGP), CNRS, Université de Lorraine, Nancy, France

In order to ensure proper and optimal extrapolation of a bioreactor process, the engineer must consider a number of technological choices. These choices include the sequence of upstream and downstream unit operations, the selection of the necessary equipment and the operating modes associated with this equipment. In terms of the biotransformation operation itself under the effect of different microorganisms, technological extrapolation decisions involve the composition of the medium, which varies according to the microorganism used, the bioreactor technology and its implementation.

With regard to the composition of the medium, which must meet the nutritional requirements of the strain used, the choice involves selecting the nutrients and determining their appropriate concentrations. Indeed, the composition of the medium has a significant influence on the course of the process, as well as on the final concentrations of the obtained products. Moreover, it can also have an impact on other process steps, including sterilization, extraction and purification. The composition of the medium can also have a significant impact on the cost of the process. In the case of large-scale metabolite production, for example, the cost of the medium can account for more than half the total cost of production.

The elements to be optimized in bioreactor culture media are the carbon source, typically sugars or mineral carbon for microalgae, the nitrogen source, which can be ammonia, nitrates, amino acids and peptides, the phosphorus source, usually phosphates, growth factors or micronutrients such as vitamins, iron, zinc, copper and manganese, as well as the source and brightness in the case of microalgae cultivation.

For bacterial, yeast and fungal cultures, most industrial fermentation media are complex media containing elements such as yeast extract, corn steep or peptones, the composition of which is not precisely known. However, media in which the concentrations of all of the elements are specified are also used. Although these media are generally more expensive, they have the advantage of facilitating product purification and ensuring reproducibility.

Microalgae can be grown under autotrophic conditions, using sunlight or artificial light as an energy source and CO_2 or bicarbonate as a carbon source. In heterotrophic cultures, microalgae use simple sugars as a carbon and energy source. The culture medium must also provide sufficient quantities of mineral ions, and possibly growth factors or vitamins.

With regard to bioreactor technology, there is a wide range of bioreactors available for microbial fermentations or microalgae cultivation. The most common technology revolves around using cells in suspension in mixed tanks. In the laboratory, microorganisms are cultivated in containers that are agitated and in small mixed bioreactors, with capacities ranging from a few liters to several tens of liters. These cultures are then extrapolated to tanks with volumes of up to several thousand cubic meters. Other, less frequently used technologies include membrane bioreactors, in which the cells are retained by microfiltration membranes, and fixed-bed or fluidized-bed bioreactors, in which the cultured cells are located on the surface or inside solid supports.

An essential feature of a bioreactor is its agitation and/or aeration system. For anaerobic microbial fermentations, tanks may be equipped with moderate agitation systems, or may not require agitation at all. In such cases, agitation of the medium can be achieved by the release of CO_2, like in the cylindrical-conical tanks used in the brewing industry. For aerobic fermentations, bioreactors are equipped with more powerful agitation and aeration systems. The mechanically aerated and agitated bioreactor, fitted with various types of agitator, is the most commonly chosen option for bacterial and yeast culture. Other, less shearing systems are used for microalgae and fungi. Other aeration configurations include bubble columns, air lifts and external loop bioreactors, where the medium is continuously stirred by rising air bubbles. The choice of agitation and aeration technology depends on the oxygen demand of the process under study, the rheological properties of the medium and the sensitivity of the cells to shear forces.

In terms of the bioreactor operating mode, aerobic or anaerobic bioreactors with suspension cells can be operated in three different modes: batch, fed-batch and continuous. In batch mode, the entire culture medium is initially introduced into the reactor. Fed-batch mode involves the initial introduction of part of the culture medium, followed by the continuous addition of the rest of the medium over time. The continuous mode is characterized by the continuous supply of fresh medium and the simultaneous removal of used medium. Continuous culture can also be implemented using several bioreactors that are arranged in a series.

The continuous bioreactor can also be combined with cell recovery and recycling. This operation can be carried out using a centrifuge, a microfiltration membrane module or simply by decantation in the case of flocculating cells.

For a specified operating mode, the bioprocess engineer must also optimize its operational variables. The predominant operational variable is the composition of the medium, which is characterized by the concentrations of substrates and nutrients added, notably sugar concentration for bacteria and minerals, and CO_2 for microalgae. Concentrations of other nutrients, such as sources of nitrogen or phosphorus, are also adjusted. The substrate concentrations to be optimized are the concentrations in the initially added medium and, in the case of continuous or fed-batch operation, the substrate concentrations in the supply medium.

The second category of operational variables concerns bioreactor aeration. It includes air flow rate and agitation power of the medium in the case of mechanically agitated biotransformation. For yeast and bacterial cultures, it is also possible to increase the mole fraction of oxygen in the inlet air in order to overcome any oxygen limitations.

Temperature and pH are other operational parameters to consider. In most cases, temperature and pH are kept constant. However, they can also be modified over time, for example, to promote metabolite excretion.

The choice of technology and operating mode for a bioreactor depends on a number of criteria. First, it is influenced by the nature of the transformation, depending on the microorganism used, whether cell or metabolite production is involved, as well as its aerobic or anaerobic nature. Second, the kinetics of the microorganism used play a crucial role, particularly with regard to inhibitions of cell growth at high concentrations of substrates and metabolites. It is also important to consider whether or not metabolite production is linked to cell growth, as well as the morphology of cultured cells and their sensitivity to shear stress in bioreactors. Additional criteria to be considered include the quality of the products obtained, the sterility requirements of the process and the economic aspects related to the cultivation process, which depend on the species of microorganism used.

Given the wide range of possible transformations and the different types of cells used (bacteria, yeast, fungi and microalgae), there is no global optimal bioreactor process design. Bioreactor technology and implementation must be specifically adapted to the biotransformation reaction, the characteristics of the cells and the industrial context in which they are used. However, to speed up the implementation and optimization of bioreactors, regardless of the strain used, the engineer can benefit from the methodology of bioprocess engineering, which favors the use of simulation models. The aim of these simulators is to predict and compare the progress of a bioprocess under different batch, continuous or fed-batch operating conditions, and with different operational variables, such as substrate concentrations or the intensity and quality of light for microalgae.

This approach to digital extrapolation and optimization of biotransformation in a bioreactor uses a "digital twin", enabling virtual biotransformations to be carried out, in order to determine the optimal operating mode and associated variables. The use of bioreactor simulators reduces the need for experimentation during the R&D phase. It also speeds up process extrapolation and reduces R&D costs.

The most effective simulators are those that rely on knowledge-based models, which are themselves based on a thorough understanding of the phenomena and factors that influence or control bioreactor operation under various operational conditions.

Thus, the *first step in the methodology for building* mechanistic or knowledge-based simulation models involves identifying the phenomena that influence bioreactor behavior or limit its operation. These phenomena include the transformation reaction by the microorganism used, material and heat transfers, as well as flows inside the bioreactor.

For a given microorganism culture operation in a bioreactor, it is the transformation reaction, that is specific to the microorganism used, that largely controls variations in substrate, cell and metabolite concentrations. It is therefore essential to understand the factors related to the composition of the medium that influence the rates of cell growth, substrate consumption and metabolite production.

During aerobic fermentation of yeast or bacteria, the maximum achievable cell concentration is often limited by the transfer of oxygen between the air and the medium. The speed of this oxygen transfer depends on two key factors: the solubility of oxygen in the liquid medium and the intensity of matter transfer. In the case of reactions involving microorganisms that are sensitive to carbon dioxide concentration or requiring CO_2, the solubility and transfer of CO_2 in the medium can also exert a significant influence on biotransformation kinetics.

A third potentially limiting phenomenon is the heat transfer between the medium and the cooling system used to regulate temperature. At high cell concentrations, the rates of heat production by the metabolism of the biological system used can reach very high levels, approaching the maximum rate of heat removal, which is controlled by heat transfer from the medium to the cooling fluid.

The fourth phenomenon to consider is the flow of the medium inside the bioreactor, which is closely linked to the rheological properties of the medium, as well as to the shape and agitation system of the bioreactor. The internal flows of the medium have a direct impact on the transfer rates of oxygen, carbon dioxide and heat within the bioreactor. In addition, the shear stresses generated by the flows can also affect cell viability.

This analysis of the fundamental processes influencing or limiting the operation of the bioreactor is particularly appropriate for its extrapolation. As part of the extrapolation process, it is important to identify the scale factors that may modify the flow, transfer and reaction processes between the laboratory bioreactor and the industrial bioreactor. Situations may arise where the microorganism's reaction plays a predominant role in the process at the laboratory scale, while oxygen and/or carbon dioxide and heat transfer become limiting factors for productivity at the industrial scale.

The second stage of the simulation model-building methodology involves mathematical modeling of the bioreactor. This phase consists of establishing the equations that describe the changes in substrate, cell biomass and metabolite concentrations during biotransformation. This methodology is based on the concept of the "modeling triangle", which combines the material and heat balances with the thermodynamic and kinetic laws defining the influencing phenomena.

Variations in concentrations of components during biotransformation are governed by instantaneous matter balance, which expresses the principle of conservation of matter at each instant. Similarly, temperature variations are determined by instantaneous heat balance, which expresses the principle of energy conservation.

In addition to the balances, the simulation models incorporate the thermodynamic and kinetic laws that define the influence of the fundamental processes identified above. The thermodynamic laws used include those describing the solubilities of oxygen and carbon dioxide in the media, and the enthalpies of the liquid and gas phases in bioreactors. The kinetic laws of biotransformation models involve those characterizing transformation reactions by the microorganism used, material and heat transfers, and fluid flows. Ultimately, modeling involves incorporating these thermodynamic and kinetic laws into the balance equations, so as to create a system of algebraic or differential equations that simulate the process involved.

Chapter 1 describes the fundamentals of bioprocess engineering, which details material and heat balance relationships, thermodynamic laws governing oxygen and carbon dioxide solubility, and oxygen and CO_2 transfer kinetics. Chapter 2 presents the bioreactor implementation methodology, which involves the development of simulation models for bioreactors operating in batch, continuous and fed-batch modes.

1

Principles of Fermentation Engineering

Jean-Marc ENGASSER and Mohamed GHOUL
Laboratoire Réaction et Génie des Procédés (LRGP), CNRS, Université de Lorraine, Nancy, France

To optimize the use of fermenters and bioreactors in general, it is necessary to master a number of concepts related to mass and energy balances, the kinetics of microorganisms, and the transfer and dissolution of oxygen and carbon dioxide. In this chapter, all of these concepts will be presented and discussed.

1.1. Mass balances

Mass balances are the most commonly used relationships in bioprocess engineering for modeling, simulating and the sizing of unit operation equipment. Mass balances are relations that express the conservation of the principle of matter (Felder and Rousseau 2005; Doran 2013; Geankoplis et al. 2018).

There are two types of mass balance. The first is the initial–final mass balance. It expresses the conservation of the quantity of matter between the beginning and the end of an operation or process. The unit of the initial–final mass balance is the kilogram. The second is the instantaneous mass balance. It expresses the conservation of the quantity of matter at any instant during an operation or process. The unit of instantaneous mass balance is the kilogram per second.

Bioreactor Implementation in the Agro-Food Industries,
coordinated by Mohamed GHOUL.

1.1.1. *Initial–final mass balances*

A distinction is made between the initial–final total mass balance and the initial–final component balance.

1.1.1.1. *Initial–final total mass balance*

As illustrated in Figure 1.1, the initial–final mass balance first applies to the total mass in a system. Conservation of total mass implies that, during an operation or process, the total mass of material inputs is equal to the total mass of material outputs. The mass of inputs includes the initial quantity of material introduced and the quantity of material added over time. The mass of outputs includes the total quantity of the material removed over time and the quantity of the material remaining at the end of the operation or process.

Expressed in kilograms, the principle of conservation of the total quantity of material, from the beginning to the end of an operation, is represented by the equality between the quantities of inputs and outputs:

Initial quantity + Input quantity = Output quantity + Final quantity

The total mass balance can be expressed in a different way by introducing the variation in the internal quantity of material, which represents the difference between the final quantity and the initial quantity:

Input quantity = Output quantity + internal variation

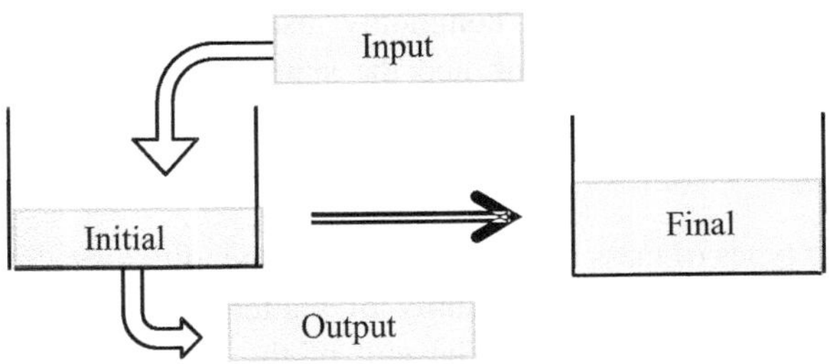

Figure 1.1. *Initial–final balance terms for total material*

1.1.1.2. *Initial–final component assessment*

The initial–final mass balance (Figure 1.2) also applies to the mass of the individual material components that are present in the system. When considering a component A, the initial–final mass balance includes the same terms of initial quantity, input quantity, output quantity and final quantity of A. If component A is involved in a chemical or biochemical reaction, where it is either produced or transformed, the mass balance includes an additional term representing the quantity produced or transformed. In the case where A is produced by the reaction, the mass balance relation for A (in kg A) is written as follows:

Initial quantity A	+	Quantity A input	+	Quantity A produced	=	Quantity A output	+	Final quantity A

If component A is transformed during the operation, the consumption of A can be considered as negative production. In this case, the initial–final balance relation is:

Initial quantity A	+	Quantity A input	–	Quantity A transformed	=	Quantity A output	+	Final quantity A

As with the total mass balance, it is possible to introduce the variation of the component's internal mass between the start and finish of the process. The component's mass balance can thus be expressed as follows:

Input quantity + quantity produced = output quantity + internal variation.

The initial–final mass balance is used to estimate the final mass of material at the end of an operation or process, taking into account the initial content and the quantities that have been added or removed over time. The balance is also used to calculate the amount of material that needs to be added or removed during an operation in order to obtain a desired final quantity. Furthermore, the initial–final component balance can be used to estimate the component mass that is either produced or transformed during a reaction based on measurements of the other terms in the balance.

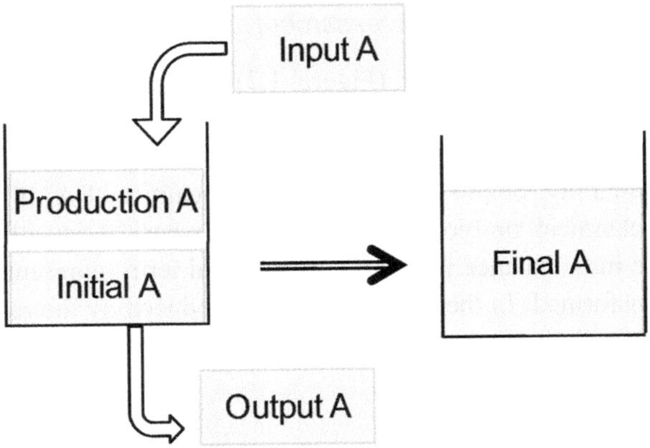

Figure 1.2. *Initial–final balance terms for component A*

1.1.2. *Instantaneous mass balances*

A distinction is made between the instantaneous total material balance and the instantaneous component balance.

1.1.2.1. *Instantaneous total mass balance*

The instantaneous mass balance (Figure 1.3) first applies to the total mass. It expresses the principle of conservation of total mass in a system at all times. At any given moment, the instantaneous total mass balance involves the three terms of rate or material flow, represented by arrows: (i) the input rate arrow indicates the material flow from the outside to the inside of the system; (ii) the output rate arrow represents the material flow from the inside to the outside of the system and (iii) the arrow indicates the rate of change of the internal quantity of material.

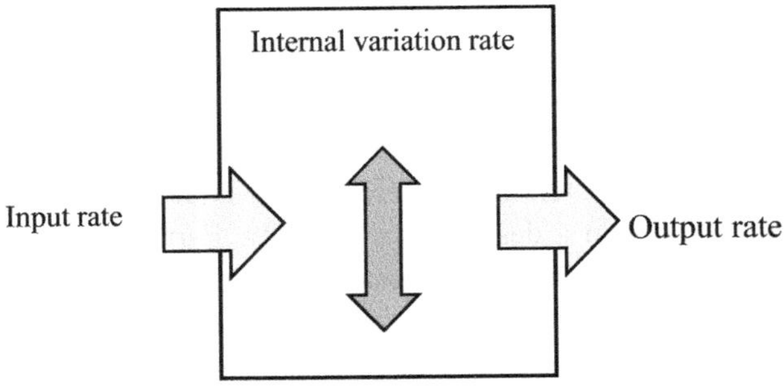

Figure 1.3. *Instantaneous balance terms for the total mass*

The instantaneous total mass balance in kg/s is as follows:

Input rate	=	Output rate	+	Rate variation

The first two terms, input rate and outlet rate, generally represent the material flow across the system boundary. The third term of the internal rate of change is equal to dm/dt, the time derivative of the total mass m inside the system. If input and output rates are equal during an operation, the rate of internal variation becomes zero. This implies that the total internal mass remains constant.

Note that the instantaneous total mass balance is equivalent to the initial–final mass balance over a very short time interval, expressed as: input = output + internal variation.

1.1.2.2. *Instantaneous component balance*

The principle of instantaneous mass balance (Figure 1.4) also applies to the various material components within the system. When there is no chemical reaction involved, the conservation of material for component A is governed by three rate or flow terms, represented by the arrows for the input rate of A into the system, the output rate of A from the system and the rate of change of the quantity of A within the system. In the presence of a chemical reaction producing or transforming component A, the instantaneous balance involves an additional arrow representing the rate at which A is produced or transformed by the reaction.

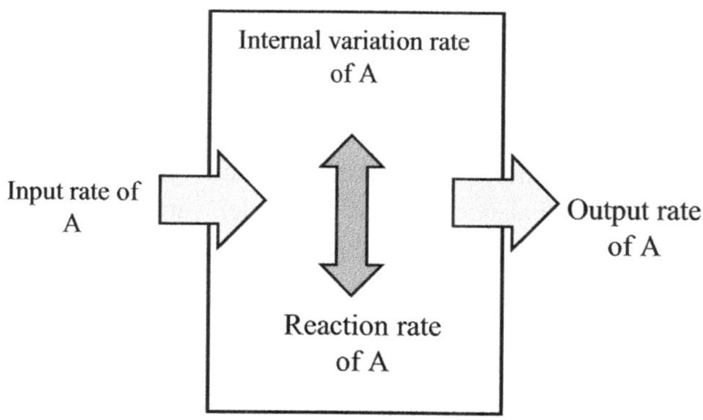

Figure 1.4. *Instantaneous balance terms for component A*

The instantaneous balance of component A, expressed in kg A/s, is as follows:

– In the case of the production of component A:

Input rate	+	Production rate	=	Output rate	+	Variation rate

– In the case of the transformation of component A:

Input rate	–	Transformation rate	=	Output rate	+	Variation rate

In these balances, the input and output rate terms represent the flow rates of component A across the system boundaries. The reaction rate term, shown as r_A, is determined by the kinetic law of the reaction. The fourth internal rate of change term is equal to dm_A/dt, the time derivative of the mass mA of A within the system. According to the principle of conservation of matter, the internal mass of A can increase or decrease according to the relative values of the input, output and reaction rates. In a steady-state system, there is no internal variation in the mass of A, and the instantaneous balance is simplified by specifying that dmA/dt is equal to zero.

It is worth noting that the instantaneous component mass balance is equivalent to the initial–final balance expressed over a very short period of time in the following form: input + production = output + internal variation.

Instantaneous mass balance is the most common relationship in bioprocess engineering. One of the main applications of instantaneous mass balance is the modeling and simulation of operations. Using total mass balance, it is possible to determine the temporal variations in the mass of material in a system. Similarly, the instantaneous component balance is used to model variations over time in the concentration of a component in a system. Another application of the instantaneous component balance is in the kinetic analysis of chemical or biochemical reactions. In a continuous or discontinuous system, the instantaneous balance of a component A enables the reaction rate of A to be determined from the measurement of the variation over time in the concentration of the component.

1.1.2.3. *Example of mass balances on a garder tank*

A mixed tank initially contains a volume V_0 of 100 L of a sugar solution with a sugar concentration S_0 of 100 g/L and a density ρ_0 of 1.04 kg/L. A feed solution, with a S_{feed} sugar concentration of 250 g/L and a density ρ_{feed} of 1.1 kg/L, is added continuously to the tank at a constant flow rate Q of 2 L/min.

Calculate the terms of the following balances:

– total mass and sugar balances on the tank, from the start to the end of a 10 min t_{feed} feeding period;

– instantaneous total material and sugar balances during tank feeding.

Initial–final total mass balance

$$V_0\ \rho_0 + Q_{feed}\ \rho_{feed} = \text{total final mass}$$

$$100\ \text{L} \times 1.04\ \text{kg/L} + 2\ \text{L/min} \times 10\ \text{min} \times 1.1 = \text{total final mass}$$

$$104\ \text{g} + 22\ \text{g} = 126\ \text{g}$$

Initial–final sugar balance

$$V_0\ S_0 + Q_{feed}\ S_{feed} = \text{final sugar mass}$$

$$100\ \text{L} \times 100\ \text{g/L} + 2\ \text{L/min} \times 10\ \text{min} \times 250\ \text{g/L} = \text{final sugar mass}$$

$$10{,}000\ \text{g} + 5{,}000\ \text{g} = 15{,}000\ \text{g}$$

Instantaneous total mass balance

$$Q\ \rho_{feed} = \text{rate variation total mass}$$

$$2\ \text{L/min} \times 1.1\ \text{kg/L} = \text{rate variation total mass}$$

$$2.2\ \text{kg/min} = 2.2\ \text{kg/min}$$

Instantaneous sugar balance

$$Q\ S_{feed} = \text{rate variation sugar mass}$$

$$2\ \text{L/min} \times 250\ \text{g/L} = \text{rate variation sugar mass}$$

$$500\ \text{g/min} = 500\ \text{g/min}$$

1.2. Heat balances

Heat is a co-product of microbial transformations. Microorganisms generate significant amounts of heat through their metabolic reactions, when converting organic substrates into biomass or metabolites.

Controlling the temperature of the fermentation medium is essential for optimal operation of a fermenter. In most cases, the temperature is kept constant throughout fermentation. However, it is also possible to have different optimal temperature ranges during the growth and metabolite production phases.

Temperature control in a fermenter is achieved by removing the heat that is generated through a cooling system, such as a cooling jacket or internal coils. To maintain a constant temperature, heat must be removed at a rate equivalent to that of its production in the medium. Removing heat from the fermentation process can be one of the difficulties in scaling up fermenters. As tank size increases, the ratio of lateral heat exchange surface area to media volume decreases, affecting the efficiency of cooling through the cooling jacket.

Modeling temperature control in a fermenter is based on the instantaneous heat balance in the plant (Figure 1.5). This balance comprises several terms: the rate of heat input into the air ($q_{a,input}$), the rate of heat generation by both agitation energy (r_{ag}) and microbial metabolism (r_f), the rate of heat output into the air ($q_{a,output}$) and by the cooling system (q_r).

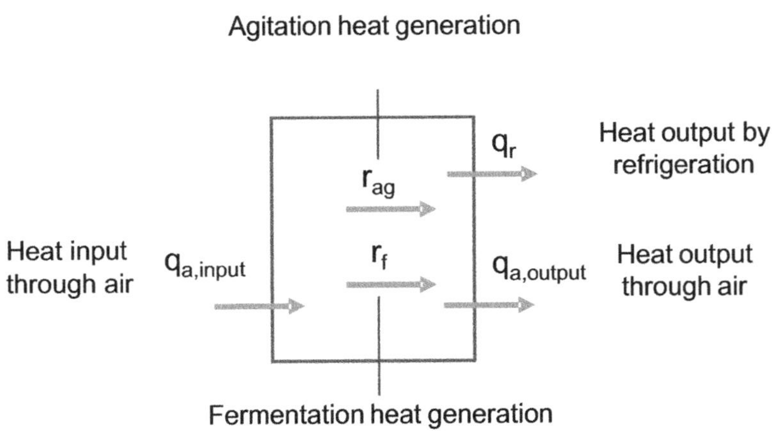

Figure 1.5. *Instantaneous heat balance terms on a fermenter. For a color version of this figure, see www.iste.co.uk/ghoul/bioreactor.zip*

In this section, we will take a detailed look at the terms of the heat balance rate in a fermenter, with the aim of better understanding how to regulate temperature during fermentation (Cooney et al 1968; Luong and Volesky 2005; Doran 2013).

1.2.1. *Heat generation rate in fermenters*

Heat is generated in the fermenter by the dissipation of mechanical agitation energy and by microbial metabolism. The heat generation rate in the fermenter, r_c (kJ/s), is the sum of the agitation and fermentation heat generation rates:

$$r_h + r_{h,ag} = r_{h,f} \qquad [1.1]$$

The agitation heat generation rate, $r_{c,ag}$ (kJ/s), is equal to the following mechanical agitation power:

$$r_{chag} = P_{ag} \qquad [1.2]$$

The agitation power P_{ag} depends on the geometry and rate of rotation of the stirrers, as well as on the rheological properties of the medium.

For aerobic fermentation, the rate of microbial heat generation $r_{c,f}$ is generally considered to be proportional to the rate of oxygen consumption in the fermentation medium:

$$r_{h,f} = Y_{\frac{c}{O2}} r_{O2} = Y_{\frac{h}{O2}} r'''_{O2} V \qquad [1.3]$$

where Yc/O_2 represents the microbial heat generated in Joules per gram of oxygen consumed. This heat depends on the microbial strain and the substrate converted. An average value is 14 kJ/g of O_2 consumed.

For anaerobic fermentations, the rate of heat generation is generally assumed to be proportional to the rate of carbon substrate consumption.

$$r_{h,f} = Y_{\frac{h}{S}} r_S = Y_{\frac{c}{S}} r'''_S \ V \qquad [1.4]$$

where Yc/S represents the microbial heat generated in J/g of substrate consumed. This heat depends on the microbial strain and the carbon substrate. For example, for anaerobic sugar fermentation by the yeast *Saccharomyces cerevisiae*, Yc/S is around 0.6 kJ/gS.

1.2.2. *Air heat input and output rates*

The heat input rate, $q_{a,input}$ (kJ/s), is equal to the air mass flow $M_{a,input}$, multiplied by the specific enthalpy of air $H_{a,input}$/kg (kJ/kg):

$$q_{a,input} = M_{a,input}\, H_{a,input/kg} \qquad [1.5]$$

The specific enthalpy of air comprises two inputs: the enthalpy of dry air, which is the product of the heat capacity C_{pa} of air by the input temperature $T_{a,input}$ air, and the enthalpy of water vapor in air, which is the product of the mass fraction of water vapor $x_{v,input}$ in the air by the specific enthalpy of water vapor. The latter depends on the heat capacity C_{pv} of the water vapor and the temperature of the input air. Thus:

$$q_{a,input} = M_{a,input}\left(C_{pa}\, T_{a,input} + x_{v,input}\left(2{,}500 + C_{pv}\, T_{a,input}\right)\right) \qquad [1.6]$$

The water vapor mass fraction $x_{v,input}$ is calculated from the relative humidity of the air at the input temperature.

Similarly, the heat output rate $q_{a,output}$ in the air flow is:

$$q_{a,output} = M_{a,output} H_{a,output/kg} \qquad [1.7]$$

At the output temperature T_{output} and water vapor mass fraction $x_{v,output}$, the heat output rate is calculated by the following relation:

$$q_{a,output} = M_{a,output}\left(C_{pa}\, T_{a,output} x_{v,output}\left(2{,}500 + C_{pv,output}\, T_{a,output}\right)\right) \qquad [1.8]$$

The output water mass fraction takes into account water evaporation inside the fermenter. Its value is highest when the output air is saturated with moisture.

1.2.3. *Cooling rate of a fermenter*

Heat is removed from the fermentation medium by a cooling jacket or internal coils through which cold water circulates (Figure 1.6).

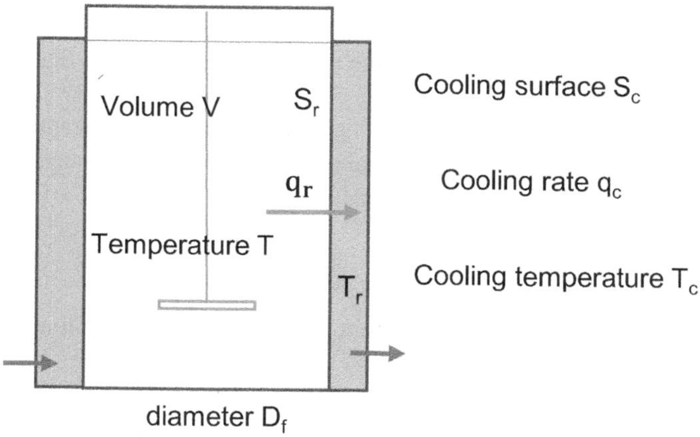

Figure 1.6. *Fermenter cooling characteristics through its cooling jacket. For a color version of this figure, see www.iste.co.uk/ghoul/bioreactor.zip*

If the cooling exchange surface is S_c and the average temperature of the circulating cooling water is T_c, the heat transfer rate q_c in J/s of the medium at temperature T is given by the kinetic law of heat transfer:

$$q_c = S_c \; h \; (T - T_c) \tag{1.9}$$

where h is the heat transfer coefficient between the medium and the cooling water.

For a cylindrical fermenter with diameter d_f and medium height h_m, the cooling jacket exchange surface is:

$$S_c = \pi d_f h_m \tag{1.10}$$

and the ratio between the exchange surface and the volume of the medium is equal to:

$$\frac{S_r}{V} = \frac{\pi \; d_f \; h_m}{\pi \;\; d_f^{\,2}/4 \;\; h_m} = \frac{4}{d_f} \tag{1.11}$$

This relation shows that the surface-to-volume ratio decreases with increasing fermenter diameter.

The heat transfer coefficient h ($J/m^2.s.°C$) is assessed from the individual thermal characteristics of the three stages involved in the heat transfer process between the medium and the cooling water, namely, h_m is the heat transfer coefficient in the fermentation medium, h_c is the heat transfer coefficient in the cooling water, l is the thickness of the fermenter wall and λ is the thermal conductivity of the steel wall.

The total resistance to heat transfer, which is the inverse of the transfer coefficient h, is determined by adding up the individual resistances of the three stages:

$$\frac{1}{h} = \frac{1}{h_m} + \frac{l}{\lambda} + \frac{1}{h_c} \qquad [1.12]$$

Thus, the overall heat transfer coefficient h is calculated by the following relation:

$$h = 1/\left(\frac{1}{h_m} + \frac{l}{\lambda} + \frac{1}{h_c}\right) \qquad [1.13]$$

1.2.4. *Heat balance of a fermenter*

The instantaneous heat balance, which expresses the principle of heat conservation at each instant, is written as a four-term balance relation, similar to the instantaneous mass balance relation. The four terms of the balance are the input rate, the output rate, the rate of production by energy conversion and the rate of change of internal enthalpy H_f. The heat balance is written as:

$$q_{a,input} + r_h = q_{a,output} + q_h + \frac{dH_f}{dt} \qquad [1.14]$$

This heat balance can initially be used to determine the rate of temperature increase in the absence of cooling of the medium. The change in internal enthalpy is then given by the following relation:

$$\frac{dH_f}{dt} = q_{a,input} + r_h - q_{a,output} \qquad [1.15]$$

The enthalpy, which represents the heat contained inside the fermenter, is essentially the enthalpy of the liquid medium. Consequently, the rate of temperature rise T in the medium is given by equation:

$$\frac{dT}{dt} = \frac{q_{a,input} + r_h - q_{a,output}}{m_m\, C_{pm}} \qquad [1.16]$$

where m_m is the mass of the medium and C_{pm} is the heat capacity of the medium.

The instantaneous heat balance is also used to assess the cooling rate required for temperature control. At a constant controlled temperature, the term for the rate of change is zero, and the cooling rate is given by the relation:

$$q_c = r_h + q_{a,input} + q_{a,output} \quad [1.17]$$

Once we know the cooling surface S_r and the heat transfer coefficient h, we can use the kinetic law expressing this cooling rate to estimate the required cooling water temperature:

$$T_c = T - \frac{q_c}{S_r\, h} \quad [1.18]$$

1.2.5. *Example: instantaneous heat balance of an aerated fermenter*

A cylindrical fermenter that is 4 m in diameter contains 100 m^3 of medium (Figure 1.7). It is mixed at an agitation power of 1 kW/m^3 and aerated by an air flow rate of 1 vvm (volume of air per volume of medium per minute). Its temperature is regulated at 35°C. When the biomass concentration (X) reaches 10 g/L and the specific oxygen consumption rate rises to 0.3 $gO_2/gX.h$, the instantaneous heat balance of the fermenter is established.

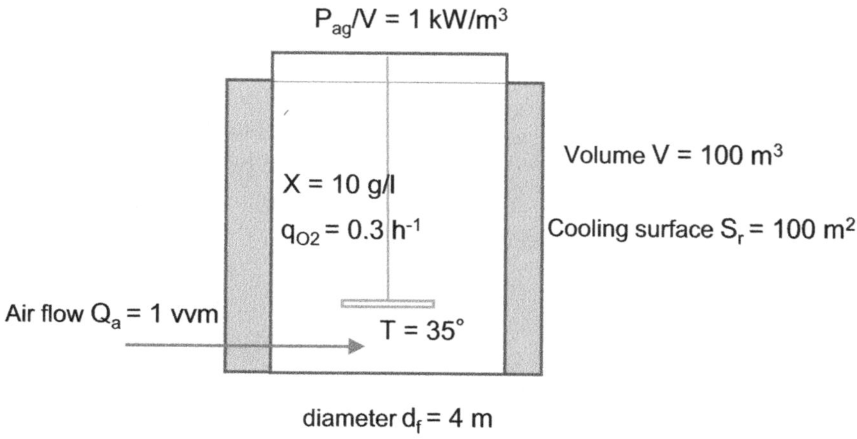

Figure 1.7. *(a) Fermenter agitation, aeration and cooling characteristics*

Calculate the rate at which the temperature rises in the absence of any cooling, and the cooling water temperature required to maintain the temperature of the medium at 35°C.

The results obtained are shown in Figure 1.7(b). The rate of heat production in the fermenter by mechanical agitation is the agitation power throughout the fermenter, in other words, 100 kJ/s. The microbial heat production rate, evaluated as the product of the volumetric rate of oxygen consumption by 14 kJ/gO_2, is equal to 1,250 kJ/kg.

The input rate of heat in the air, at a temperature of 20°C, is the air mass flow rate multiplied by the specific enthalpy of the air under the input conditions. For the water vapor mass fraction corresponding to a relative humidity of 20%, the input rate is approximately 50 kJ/s. In the output air, at a temperature of 35°C and assumed to be saturated with water vapor, the heat output rate reaches 240 kJ/s.

In the absence of a fermenter cooling system, the temperature of the medium rises at a rate of 0.15°C/min. In order to maintain the temperature of the medium at 35°C, the cooling rate required is around 1,060 kJ/s. As the lateral surface area of the fermenter is 100 m^2, for a heat transfer coefficient of around 500 W/m^2.°C, the average temperature of the cooling water circulating in the jacket must be close to 13°C.

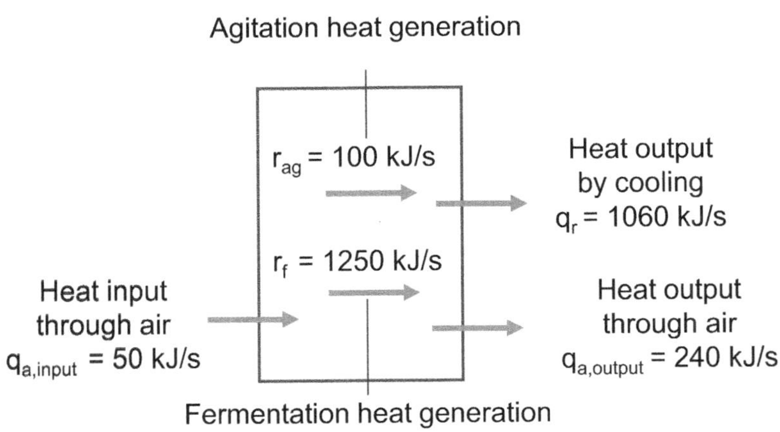

Figure 1.7. *(b) Values of the instantaneous heat balance terms for the fermenter. For a color version of this figure, see www.iste.co.uk/ghoul/bioreactor.zip*

1.3. Microbial kinetics

Microbial kinetics is the field of bioprocess engineering that deals with the kinetic laws governing the transformation reactions carried out by microbial cells. These laws put forward the relationships that express the rates of cell growth and death, nutrient consumption and metabolite production, as a function of the cell environment (Bailey and Ollis 1986; Blanch and Clark 1996; Nielsen 2006; Shuler et al. 2017; Liu 2020).

Kinetic laws represent the individual characteristics of the microbial strain used. They are essential for modeling and simulating fermentation processes. By combining these laws with instantaneous mass balance relationships, we can determine how nutrient, cell and metabolite concentrations vary over time during fermentation.

This chapter describes the stoichiometric coefficients of microbial reactions, microbial fermentation rates and their experimental assessment, as well as the most common laws of microbial kinetics.

1.3.1. *Stoichiometry of microbial reactions*

Microbial living cells function as microreactors, containing thousands of simultaneous and consecutive enzymatic reactions that constitute cellular metabolism. These reactions, each with their own stoichiometry, determine the rates at which cells utilize nutrients, grow and produce metabolites.

To model fermentations, the first step is to establish the overall stoichiometry of the microbial transformation reaction. This involves defining the stoichiometric coefficients that describe the relationships between substrate consumption, and biomass and metabolite production.

1.3.1.1. *Molar stoichiometric coefficients*

Like chemical reactions, stoichiometric coefficients can be expressed on a molar basis. For example, cell growth on sugar under aerobic conditions can be represented using the following stoichiometric relation:

$$\text{a Sugar} + \text{b } O_2 \rightarrow \text{c Cells} + \text{d Metabolite} + \text{e } CO_2 \qquad [1.19]$$

a, b, c, d and e represent the molar stoichiometric coefficients for the various components involved.

To determine the molar stoichiometric coefficients, it is necessary to know the molecular composition of the species involved in the reaction, and in particular that of the cell biomass. An approximate molecular composition of the biomass, such as $C_5H_7NO_2$, can be assumed. However, biomass composition can change during fermentation.

1.3.1.2. *Mass stoichiometric coefficients*

Due to the complexity of microbial processes, a more common approach to microbial kinetics is the use of stoichiometric mass coefficients. These coefficients are the proportionality factors that relate the masses of the substrates consumed to those of the products.

The stoichiometric mass coefficient, $Y_{A/B}$, represents the mass of compound A consumed or produced per gram of compound B produced or consumed.

In the case of aerobic cell growth on sugar, for example, microbial transformation can be represented by the following stoichiometric relation:

$$Sugar + Y_{O2/X}\,O_2 \rightarrow Y_{X/S}\,\text{Cells} + Y_{M/S}\,\text{Metabolite} + Y_{CO2/XO2}\,CO_2 \quad [1.20]$$

The stoichiometric mass coefficients are $Y_{O2/X}$ in g of oxygen consumed per gram of biomass produced, $Y_{X/S}$ in g of biomass produced per gram of sugar consumed, $Y_{M/S}$ in g of metabolite produced per gram of sugar consumed and $Y_{CO2/O2}$ in g of CO_2 produced per gram of O_2 consumed.

Stoichiometric coefficients frequently vary during fermentations, reflecting changes in cellular metabolism. However, there are generally phases of fermentation during which the stoichiometries remain relatively constant. One example is the cell growth phase in the absence of nutritional restrictions.

1.3.1.3. *Example: stoichiometry of Corynebacterial growth*

A Corynebacterium is grown aerobically on sugar in a batch fermenter. The medium contains an initial sugar concentration S_0 and is seeded with a cell concentration X_0. Sugar concentration S and biomass concentration X are measured during fermentation (Figure 1.8). The stoichiometric coefficient of bacterial growth is determined.

The stoichiometric coefficient $Y_{X/S}$ in g of biomass formed per gram of sugar consumed, also known as the conversion efficiency from sugar to biomass, is determined graphically by plotting the concentration of biomass produced,

X – X_0, against the concentration of sugar consumed, S_0 – S, at different points in the fermentation process.

The resulting curve is more or less a straight line. Consequently, the stoichiometric coefficient $Y_{X/S}$ is constant during bacterial growth, equal to 0.46 g of biomass per gram of sugar, corresponding to the slope of the straight line.

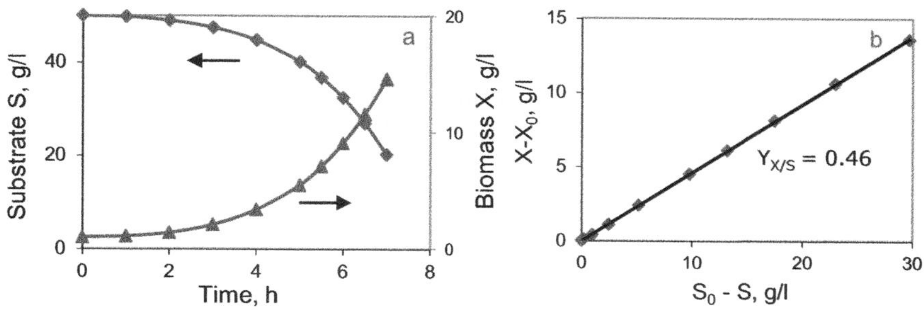

Figure 1.8. *(a) Variation in substrate S and biomass X concentrations during Corynebacterial growth. (b) Biomass formed X-X₀ as a function of substrate consumed to determine the stoichiometric coefficient* $Y_{X/S}$

1.3.2. *Fermentation rates*

1.3.2.1. *Definition of fermentation rates*

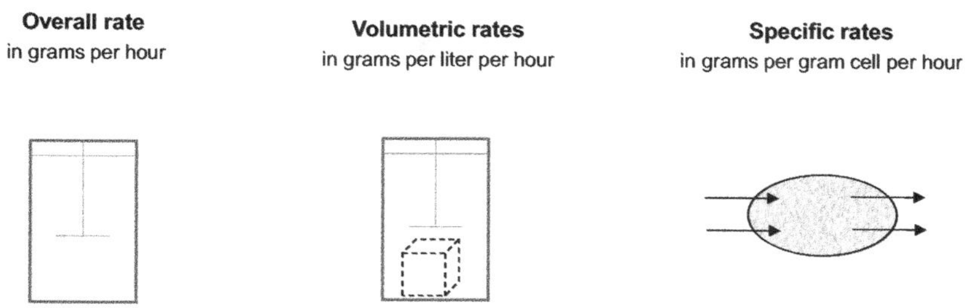

Figure 1.9. *Overall, volumetric and specific cell rates in a fermenter*

To model fermentation processes, various quantities are used to describe the rates of cell growth and death, substrate consumption and metabolite production (Figure 1.9).

The first variables are the overall rates for the entire medium in the fermenter. The units of overall rates are grams per hour of substrate, biomass or metabolite.

The second variables are the volumetric rates per unit volume of medium, expressed in grams per liter per hour. If the medium is homogeneous, the volumetric rate is equal to the overall rate divided by the volume of the medium.

Kinetic laws are generally expressed in terms of the third-rate variable, specific rates per unit cell quantity, usually per gram of dry cell weight. Specific rates are obtained by dividing volumetric rates by cell biomass concentration.

Overall, volumetric and specific rates are used to quantify the processes of cell growth and death, substrate consumption and metabolite production. The corresponding notations are shown in Figure 1.10.

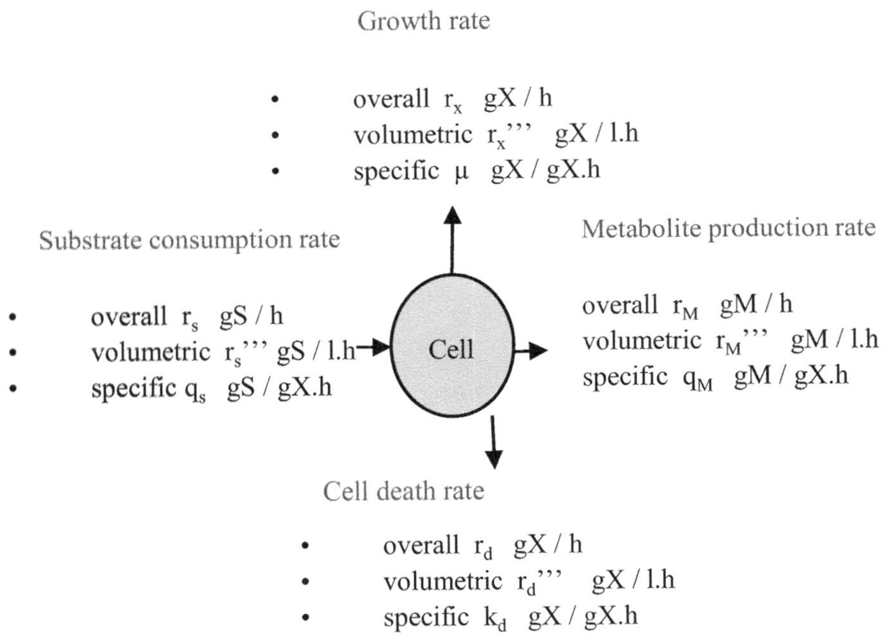

Figure 1.10. *Defining the rates of growth, cell death, substrate consumption and metabolite production in a fermenter. For a color version of this figure, see www.iste.co.uk/ghoul/bioreactor.zip*

1.3.2.2. *Experimental determination of batch fermentation rates*

In batch fermentation, the rates of growth, nutrient consumption and metabolite production change over time. These variations in fermentation rates can be determined by measuring nutrient, biomass and metabolite concentrations throughout fermentation, and applying instantaneous mass balance relationships to the compounds involved.

1.3.2.2.1. Determination of growth rates

From the measured biomass concentration X (g/L), the overall cell production rate r_X is determined by the instantaneous biomass balance in the fermenter (gX/h):

$$rx = \frac{dmx}{dt} \quad [1.21]$$

with m_X being the mass of biomass in the fermenter.

By introducing the volume of the medium V, the biomass balance becomes:

$$Vrx''' = \frac{Vdx}{dt} \quad [1.22]$$

or

$$rX''' = \frac{dx}{dt} \quad [1.23]$$

Consequently, at any given time, the growth rate rX‴ (gX/L.h) is equal to the dX/dt derivative of the biomass concentration.

The specific growth rate μ (h^{-1}) is calculated by dividing the volumetric rate rX‴ by the biomass concentration X.

Similar equations can be used to determine the overall, volumetric and specific rates of metabolite production.

1.3.2.2.2. Determination of substrate consumption rates

From measurements of substrate concentrations during batch fermentation, the overall rate $r_{S'''}$ of substrate consumption is determined using the instantaneous substrate balance of the fermenter (gS/h):

$$-rs''' = \frac{dms}{dt} \quad [1.24]$$

with m_S being the mass of substrate in the fermenter.

By introducing the volume of the medium V, the substrate balance becomes:

$$-Vrs''' = \frac{VdS}{dt} \qquad [1.25]$$

or

$$-rs''' = \frac{dS}{dt} \qquad [1.26]$$

Thus, at any given time, the substrate consumption rate rS''' (gS/L.h) is equal to minus the dS/dt derivative of the substrate concentration.

The specific substrate consumption rate q_S (gS/gX.h) is calculated by dividing the volume rate r_S''' by the biomass concentration X.

1.3.2.2.3. Example: Corynebacteria production rates in batch fermentation

A Corynebacterium is produced on glucose in a batch fermenter. The medium contains an initial concentration of 50 g/L of glucose and is seeded with a concentration of 1 g/L of cells. Glucose and biomass concentrations are measured during fermentation (Figures 1.11(a) and (b)). The volumetric and specific growth rates during fermentation are determined.

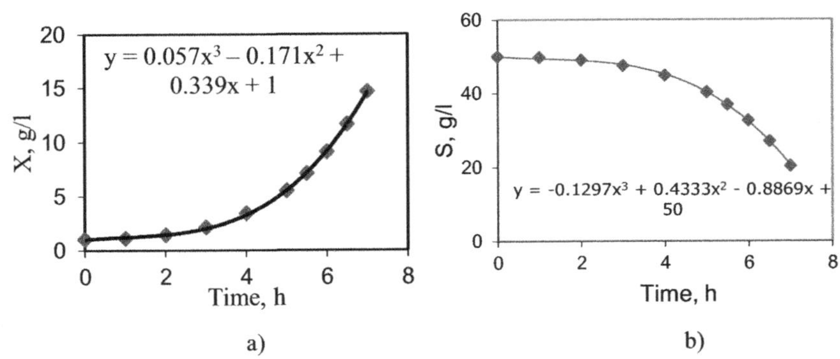

Figure 1.11. *(a) Corynebacteria X biomass variation over time. (b) Substrate S variation over time*

To assess the volumetric growth rate r_X''' of bacteria, the derivative dX/dt of biomass concentration is determined numerically on a spreadsheet using the trend curve feature.

The experimental points are modeled by a third-degree polynomial:

$$X = at^3 + bt^2 + c\,t + d \qquad [1.27]$$

The four constants a, b, c and d are identified on the graph.

The growth velocity r_X'', equal to the derivative dX/dt, is calculated by the relation:

$$rX''' = \frac{dX}{dt} = 3at^2 + 2bt + c \qquad [1.28]$$

and the specific growth rate μ in h^{-1} is calculated by dividing the volumetric rate r_X''' by the biomass concentration (Figures 1.11(c) and (d)).

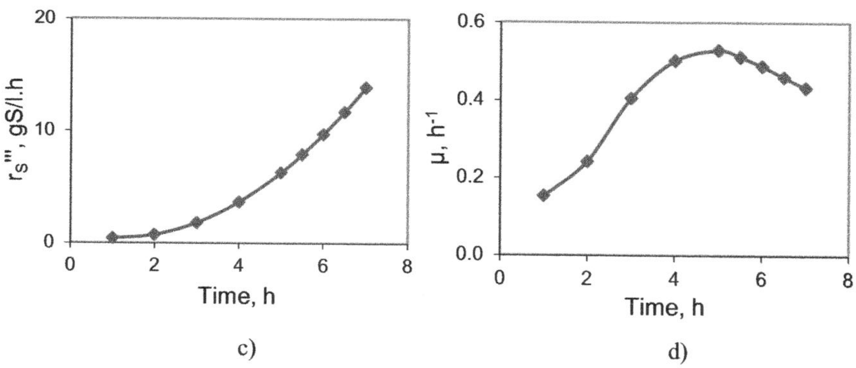

Figure 1.11. *(c) Variation in the volumetric growth rate over time. (d) Variation in the specific growth rate over time*

According to the results obtained, volumetric and specific growth rates show distinct temporal variations. The volumetric growth rate increases throughout fermentation, following the increase in cell concentration. The specific growth rate, on the other hand, only increases during the initial latent phase. It reaches a maximum value, then begins to decrease, probably due to oxygen limitations in the medium.

A similar numerical method is used for sugar consumption rates (Figures 1.11(e) and (f)).

The decrease in sugar concentration over time is modeled by a third-degree polynomial:

$$S = at^3 + bt^2 + c\,t + d \qquad [1.29]$$

and the volumetric sugar consumption rate is calculated by the relation:

$$rS''' - \frac{dS}{dt} = -\left(3at^2 + 2bt + c\right) \qquad [1.30]$$

The specific sugar consumption rate q_S is calculated by dividing the volumetric rate by the biomass concentration.

Both volumetric and specific sugar consumption rates show similar temporal variations to the corresponding bacterial growth rates. This similarity is explained by the constant conversion efficiency of sugar into bacteria during fermentation.

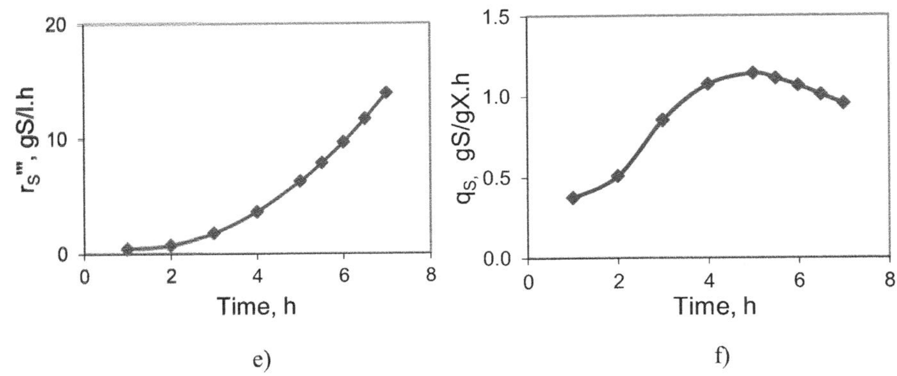

Figure 1.11. *(e) Variation in the volumetric rate of substrate consumption over time. (f) Variation in the specific rate of substrate consumption over time*

1.3.2.3. *Experimental determination of rates in continuous fermentation*

Continuous fermentations are used to assess cell growth and metabolism rates under substrate-limited conditions. During a continuous fermentation, the fermenter is fed continuously at a flow rate Q containing a concentration of substrate S_{feed}. The medium is withdrawn at the same rate to maintain a constant volume V in the fermenter.

Continuous fermentations generally present two kinetic phases: an initial dynamic phase characterized by varying substrate and biomass concentrations, followed by a stationary phase where substrate and biomass concentrations remain constant (Figure 1.12).

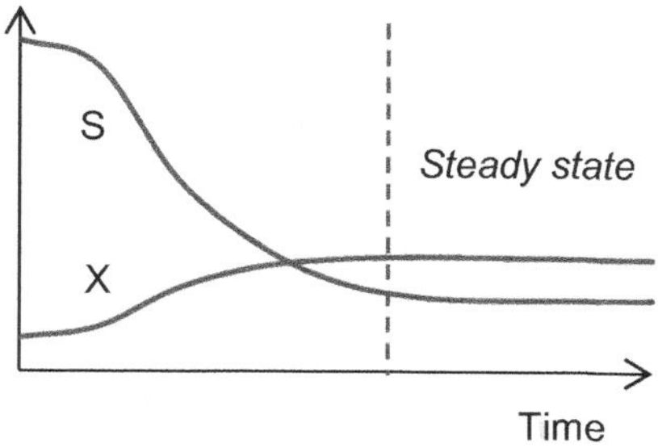

Figure 1.12. *Variation in substrate S and biomass X concentrations during continuous fermentation*

Temporal variations in fermentation rates are determined from measured concentrations of nutrients, biomass and metabolites during fermentation. Instantaneous mass balances of compounds are again used to calculate different fermentation rates.

1.3.2.3.1. Determination of growth rates

The overall biomass production rate r_X is determined by the instantaneous biomass balance of the fermenter:

$$rx''' = \mathrm{QX} + \frac{dX}{dt} \qquad [1.31]$$

Compared to the biomass balance of the batch fermenter, the continuous reactor balance involves an additional output rate term, equal to the flow rate Q multiplied by the biomass output concentration X, which is equal to the biomass concentration inside the well-mixed medium.

By introducing the volume of the medium V, the biomass balance becomes:

$$Vrx''' = \mathrm{QX} + \frac{VdX}{dt} \qquad [1.32]$$

Consequently, at each instant, the volumetric growth rate is calculated as:

$$rx''' = \frac{\mathrm{QX}}{\mathrm{V}} + \frac{dX}{dt} \qquad [1.33]$$

When the steady-state is reached, the biomass concentration is constant, and the equation for the volumetric growth rate is written as:

$$rx''' = \frac{QX}{V} \quad [1.34]$$

When expressing the volumetric rate as a function of the specific growth rate, the steady-state biomass balance becomes:

$$\mu X = \frac{QX}{V} \quad [1.35]$$

or

$$\mu = \frac{Q}{V} = D \quad [1.36]$$

where D is the dilution ratio.

Thus, when the steady-state is reached, the specific growth rate μ is simply equal to the fermenter dilution ratio D, which is the ratio between the flow rate Q of medium and the internal volume V of medium.

1.3.2.3.2. Determination of substrate consumption rates

The overall rate r_S of substrate consumption is determined by the instantaneous substrate balance of the continuous fermenter:

$$QS_{feed} - r_s = QS + \frac{dm_S}{dt} \quad [1.37]$$

Compared to the substrate balance of the batch fermenter, the balance of the continuous fermenter involves an additional input rate term, equal to the flow rate multiplied by the substrate feed concentration, and an output rate term, equal to the flow rate multiplied by the substrate concentration in the fermenter.

The volumetric rate r_S''' of consumption is assessed by the following relation:

$$QS_{feed} - Vr_s = QS + V\frac{dS}{dt} \quad [1.38]$$

or

$$r_s''' = \frac{Q(S_{feed} - S)}{V} - \frac{dS}{dt} \quad [1.39]$$

In a steady state, the substrate concentration in the fermenter reaches a constant value and the equation for r_S''' becomes:

$$rs''' = D(S_{\text{feed}} - S) \quad [1.40]$$

Under these conditions, the specific rate q_S of substrate consumption is calculated by the following relation:

$$q_s = \frac{D(S_{\text{feed}} - S)}{X} \quad [1.41]$$

1.3.3. *Microbial kinetic laws*

The laws of microbial kinetics are the relations that predict cell transformation rates according to various environmental parameters. Many factors can influence these rates (Figure 1.13). These initially include the composition of the fermentation medium, notably the concentrations of substrates consumed and the metabolites excreted. Physicochemical variables such as temperature, pH and shear stress can also have a significant impact on microbial rates.

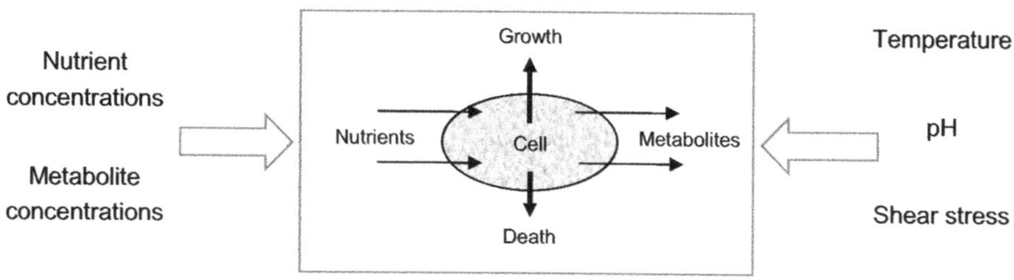

Figure 1.13. *Factors influencing cell rates in a fermenter*

During fermentation, the rates of cell growth and metabolism vary with changes in the composition of the medium. As the processes of nutrient consumption, cell growth and metabolite production are closely linked, their rates are not independent of one another. Stoichiometric relationships exist between the rates of nutrient consumption, growth and metabolite production.

For kinetic modeling of microbial fermentation, the kinetic laws are generally established in the order shown in Figure 1.14.

First, the specific cell growth rate is expressed as a function of medium composition, including substrate and metabolite concentrations. Next, the specific rates of metabolite production are established as a function of the specific rate of cell growth and/or the concentrations of substrates and metabolites in the medium. Lastly, the specific rates of substrate consumption are directly related to the preceding rates of cell growth and metabolite production. When there is significant cell death during fermentation, its specific rate may be expressed as a function of medium composition or physicochemical parameters.

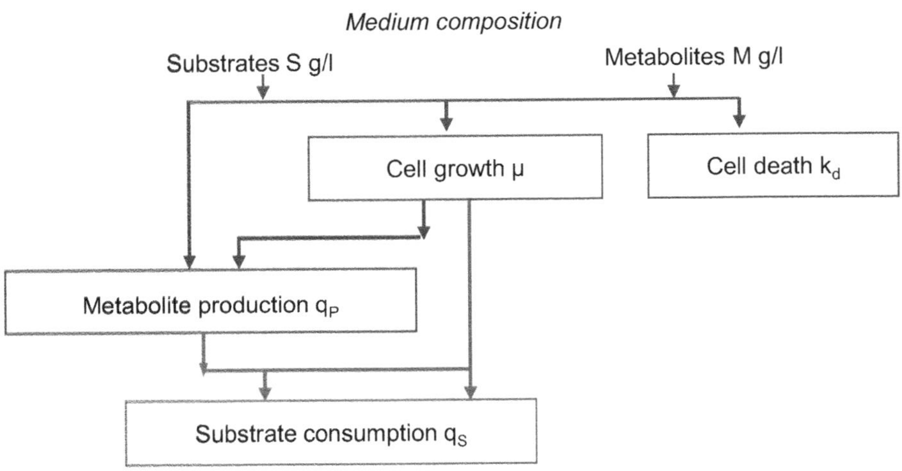

Figure 1.14. *Relation between specific cell rates and composition of the medium. For a color version of this figure, see www.iste.co.uk/ghoul/bioreactor.zip*

1.3.3.1. *Cell growth kinetics*

The specific rate of microbial growth μ is generally determined as the product of the maximum growth rate μ_{max} and growth-limiting factors due to reduced substrate concentrations and/or the accumulation of inhibitory metabolites in the medium.

Under optimal conditions, when limitations related to substrates or inhibitory metabolites are absent, cells grow at their maximum specific growth rate μ_{max}, measured in grams of cells produced per gram of existing cells per hour. For industrial microorganisms, μ_{max} is generally between 0.1 and 0.6 h^{-1}, corresponding to doubling times of between 1 h and 7 h.

Growth limitation by a substrate refers to the reduction in specific growth rate when the substrate concentration falls below a certain threshold. Similarly, growth limitation by an inhibitory metabolite is the reduction in specific growth rate when the concentration of the metabolite becomes too high.

1.3.3.1.1. Monod's law of growth

The limitation of specific growth rate due to substrate concentration is commonly described by Monod's kinetic law (Figure 1.15(a)).

This law establishes the mathematical relationship between the specific growth rate μ (h^{-1}) and the substrate concentration S:

$$\mu = \frac{\mu_{max}\ S}{K_S + S} \qquad [1.42]$$

μ_{max} represents the maximum specific growth rate at high substrate concentrations, while the K_S constant, expressed in g/L, corresponds to a substrate concentration with a growth rate that has halved.

A substrate is considered kinetically limiting when its concentration is approximately less than 10 times the value of K_S. By increasing the value of the K_S constant in Monod's law, the limiting effect of the substrate is pushed to higher concentrations (Figure 1.15(b)).

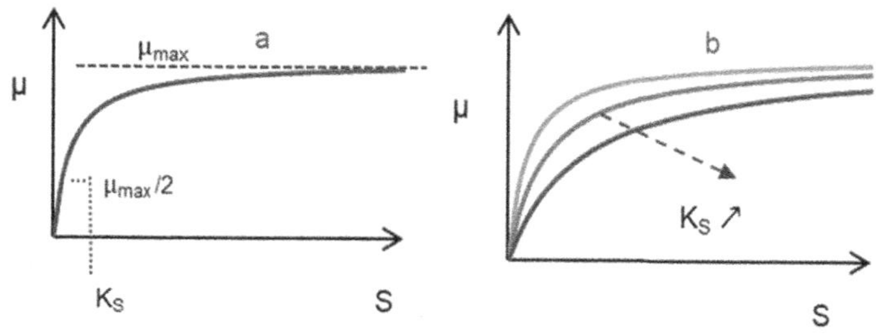

Figure 1.15. *(a) Illustration of Monod's kinetic law for specific growth rate μ. (b) Influence of K_S constant on μ as a function of substrate S concentration*

In microbial fermentations, the specific growth rate can be simultaneously limited by several substrates such as sugar, oxygen or a nitrogenous substrate. In such cases, Monod's law can be generalized by introducing several limiting terms, each characterized by the K_S constant of the substrate under consideration:

$$\mu = \mu_{max} \left(\frac{S_1}{K_{S1} + S_1}\right)\left(\frac{S_2}{K_{S2} + S_2}\right) \qquad [1.43]$$

1.3.3.1.2. Example: identification of Monod's law of growth

The specific growth rate μ, of a bacterium is determined in a continuous fermenter at different sugar concentrations. Experimental results show that the specific speed μ of the bacteria increases with the sugar concentration S, reaching a maximum value at high sugar concentrations (Figure 1.16(a)).

The applicability of Monod's kinetic law is verified from the linearized representation of 1/μ as a function of 1/S:

$$\frac{1}{\mu} = \frac{K_S + S}{\mu_{max} S} = \frac{1}{\mu_{max}} + \frac{K_S}{\mu_{max}} \frac{1}{S} \qquad [1.44]$$

On this graph, the experimental points are aligned for the bacteria under study and the two kinetic parameters μ_{max} and K_S are assessed from the slope and intersection of the line with the y-axis: μ_{max} is equal to 0.58 h^{-1} and K_S is close to 1 g/L (Figure 1.16(b)).

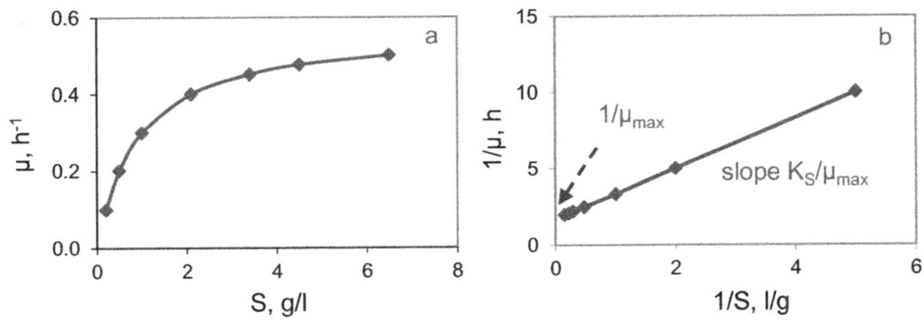

Figure 1.16. *(a) Specific growth rate of Corynebacteria μ as a function of sugar concentration S. (b) Representation of 1/μ as a function of 1/S to determine the kinetic constants μmax and Ks of Monod's law*

1.3.3.1.3. Law of growth with substrate excess inhibition

At high substrate concentrations, for example, > 100 g/L for a sugar, inhibition of microbial growth due to substrate excess can be observed. Under such conditions, the specific growth rate μ is no longer equal to μ_{max}, but progressively decreases with increasing substrate concentration.

To describe this phenomenon of substrate inhibition, the kinetic law often used is the modified Monod's law, which includes an additional S^2/K_{Si} term in the denominator.

$$\mu = \frac{\mu_{max}\ S}{K_S + S + \frac{S^2}{K_{Si}}} \qquad [1.45]$$

The inhibition constant, K_{Si}, represents the substrate concentration at which the specific growth rate is reduced to around half of μ_{max}. By increasing the value of K_{Si} in the kinetic law, inhibition by substrate excess is pushed toward higher substrate concentrations (Figures 1.17(a) and (b)).

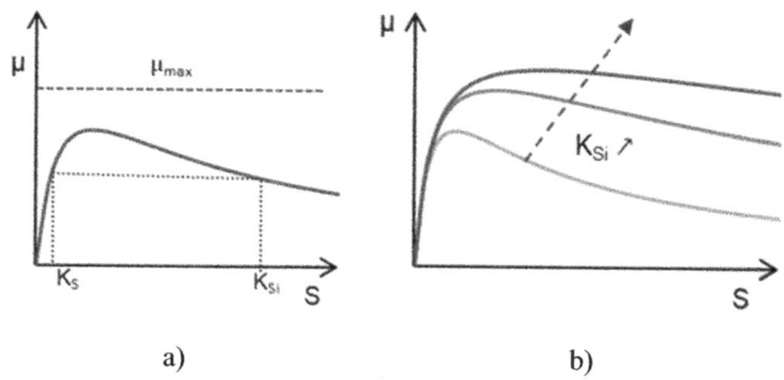

Figure 1.17. *(a) Illustration of the growth kinetic law with substrate inhibition. (b) Influence of the inhibition constant K_{Si} on the variation of the specific growth rate μ as a function of substrate S concentration*

1.3.3.1.4. Law of growth with product inhibition

The specific growth rate of a microbial cell can be reduced by an excessive concentration of an excreted metabolite. The simplest law of growth inhibition by a metabolite is linear. In this case, Monod's kinetic law is supplemented by an inhibition term, depending on the concentration of metabolite M (Figure 1.18(a)):

$$\mu = \mu_{max}\left(\frac{S}{K_S + S}\right)\left(1 - \frac{M}{M_m}\right) \qquad [1.46]$$

M_m represents the maximum metabolite concentration at which the growth rate becomes zero. Using this linear inhibition law, the inhibitory effect of the metabolite is reduced by increasing the maximum metabolite concentration M_m. Nonlinear types of inhibition can also be observed and represented by a kinetic law in the following form (Figure 1.18(b)):

$$\mu = \mu_{max}\left(\frac{S}{K_S + S}\right)\left(1 - \frac{M}{M_m}\right)^{\alpha} \quad [1.47]$$

In this form, the inhibition curve, which represents the specific growth rate μ as a function of metabolite M concentration, is concave downwards for α values below 1 and concave upwards for α values above 1.

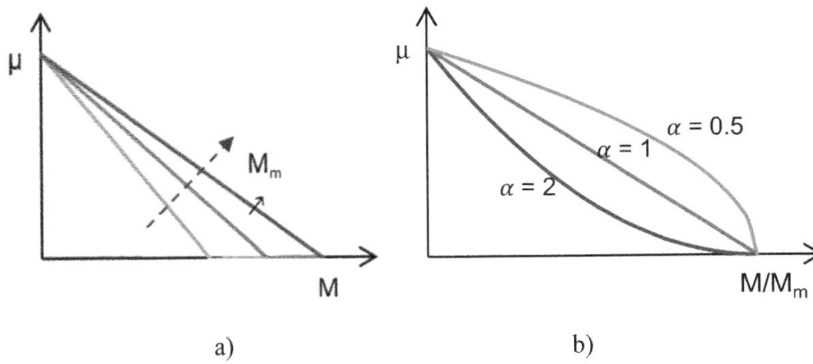

Figure 1.18. *(a) Influence of the maximum concentration of metabolite M_m on μ as a function of M in the case of linear inhibition. (b) Influence of the constant α on μ as a function of the standardized metabolite concentration M/M_m in the case of nonlinear inhibition*

1.3.3.2. *Kinetic laws of cell death*

During fermentation, the rate of denaturation or cell death is often negligible under conditions that are favorable to cell growth. However, this rate can become significant under conditions of severe growth limitation, for example, at high concentrations of inhibitory metabolites. Excessive shear stress generated by fermenter agitation can also accelerate the death of the most sensitive cells. When taking this into account in kinetic modeling, the specific rate of cell death (k_d) is generally assumed to be constant.

The most influential variable in cell death is temperature. The acceleration of the specific rate of cell death with temperature is commonly represented by Arrhenius' law:

$$k_d = A_d \exp\left(-\frac{E_{ad}}{RT}\right) \quad [1.48]$$

E_{ad} is the activation energy (in the range of 200–300 kJ/mol), A_d is the pre-exponential factor, R is the gas constant (8.32 J/mol-K) and T is the temperature in Kelvin.

Arrhenius' law of microbial denaturation is frequently used to model thermal sterilization operations.

1.3.3.3. *Kinetic laws of metabolite production*

Microorganisms are commonly used to produce metabolites or proteins. Unwanted metabolites can also be excreted during fermentation. The kinetic laws that describe the specific rates of metabolite production vary according to their association with cell growth. Indeed, some metabolites are closely linked to cell growth, while others are not. This fundamental difference in kinetic profiles has a significant impact on the production strategies and controls that need to be implemented in order to optimize the production of desired metabolites while minimizing undesirable ones.

1.3.3.3.1. Growth-associated metabolites

Several microbial metabolites are produced during the growth phase. These include ethanol, acetate and lactate, which are co-products of yeast and bacterial growth metabolism (Figure 1.19(a)).

For growth-associated metabolites, the specific metabolite production rate q_M (gM/gX.h) is directly related to the specific growth rate μ according to the following formula (Figure 1.19(b)):

$$q_M = Y_{M/X}\mu \qquad [1.49]$$

$Y_{M/X}$ represents the stoichiometric coefficient expressing the quantity in grams of metabolite excreted per gram of cell biomass produced.

In continuous or fed-batch fermentations, when microbial cells are exposed for prolonged periods to low substrate concentrations, metabolite production can decrease (Figure 1.19(c)). This is the case, for example, with ethanol production by the yeast *Saccharomyces cerevisiae* at very low sugar concentrations. This decrease can be modeled by assuming a variation in the stoichiometric coefficient $Y_{M/X}$ as a function of substrate concentration, according to the following relation:

$$Y_{M/X} = Y_{M/Xxmax}\frac{S^n}{K_M + S^n} \qquad [1.50]$$

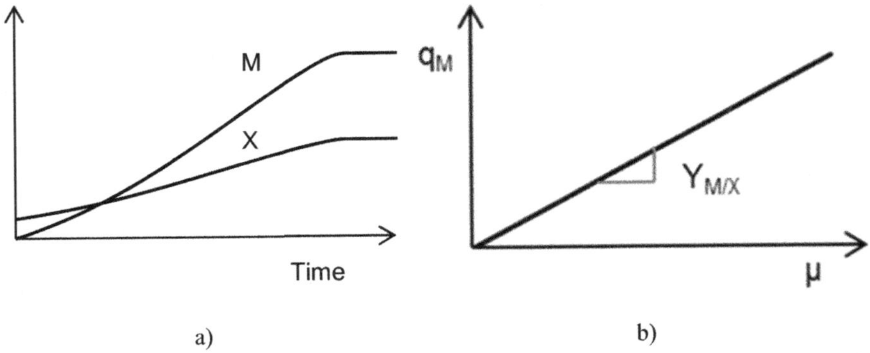

Figure 1.19. *(a) variation over time of the concentrations of biomass X and metabolite M. (b) Variation of q_M as a function of μ for a constant $Y_{M/X}$ stoichiometric coefficient*

$Y_{M/Xmax}$ represents the maximum value of the stoichiometric coefficient observed at high substrate concentrations and K_M is a constant. For exponent n values greater than 1, the $Y_{M/X}$ variation curve takes on a sigmoidal form. In this case, there is a threshold substrate concentration, below which metabolite excretion becomes negligible.

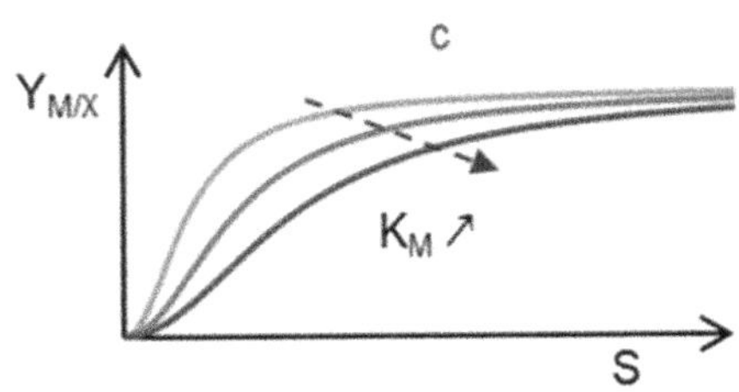

Figure 1.19. *(c) Influence of the kinetic constant K_M on the variation of the stoichiometric coefficient $Y_{M/X}$ as a function of the substrate concentration S*

1.3.3.3.2. Nongrowth associated metabolites

In most fermentations, the production of a metabolite is not directly linked to cell growth. Instead, it is often found that the metabolite is not excreted during the rapid growth phase. Instead, its excretion occurs during the slow growth phase or, in some cases, when growth has stopped altogether (Figures 1.20(a) and (b)).

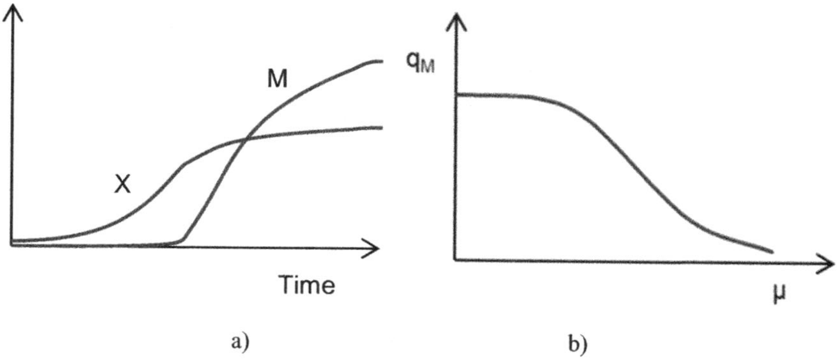

Figure 1.20. *(a) Variation in biomass X and metabolite M concentrations over time. (b) Variation of specific metabolite production rate q_M as a function of growth rate μ*

In these situations, where there is no association with growth, the kinetic law governing the specific rate of metabolite production becomes more complex. It may involve a maximum specific rate of metabolite production (q_{Mmax}) expressed in grams of metabolites per gram of cells per hour, as well as rate-limiting terms at low substrate concentrations and inhibition terms at high concentrations of metabolites excreted into the medium. An example of a kinetic law for metabolite production, comprising a rate-limiting term based on sugar concentration S and an inhibition term based on metabolite concentration M, is as follows:

$$q_M = q_{M.max}\left(\frac{S}{K_S + S}\right)\left(1 - \frac{M}{M_m}\right) \quad [1.51]$$

1.3.3.4. *Substrate consumption rate laws*

The specific substrate consumption rate q_s (gS/gX.h) is determined by a kinetic law comprising one or more terms. The general formula is as follows:

$$q_S = Y_{S/X}\,\mu + m_S + Y_{S/M}\,q_M \quad [1.52]$$

The first term corresponds to growth-related substrate consumption, which is directly proportional to the specific growth rate. The constant of proportionality is the stoichiometric coefficient $Y_{S/X}$, expressed in grams of substrate consumed per gram of biomass produced.

The second term corresponds to substrate consumption for maintenance m_S (gS/gX.h). It represents the rate of substrate consumption when growth has stopped.

In general, at high growth rates, the maintenance consumption term is significantly lower than the substrate consumption rate associated with growth.

In situations of high metabolite production, the kinetic law incorporates a third substrate consumption term, proportional to the specific metabolite production rate q_M. The constant of proportionality is the stoichiometric coefficient $Y_{S/M}$, expressed in grams of substrate consumed per gram of metabolite produced.

EXAMPLE.– Identifying the sugar consumption law of a bacterium:

The specific sugar consumption rate of a bacterium is determined in a continuous steady-state fermenter, at different specific growth rates. Identify the kinetic law of sugar consumption of the bacteria.

The experimental curve q_S as a function of the specific growth velocity μ is a straight line (Figure 1.21). The specific velocity q_S can be modeled by the following relation:

$$q_S = Y_{S/X}\,\mu + m_S \qquad [1.53]$$

From the slope and the intersection of the straight line with the y-axis, the stoichiometric coefficient $Y_{S/X}$ is equal to 2.1 gS/gX and the sugar consumption rate for maintenance m_S is equal to 0.06 gS/gX.h

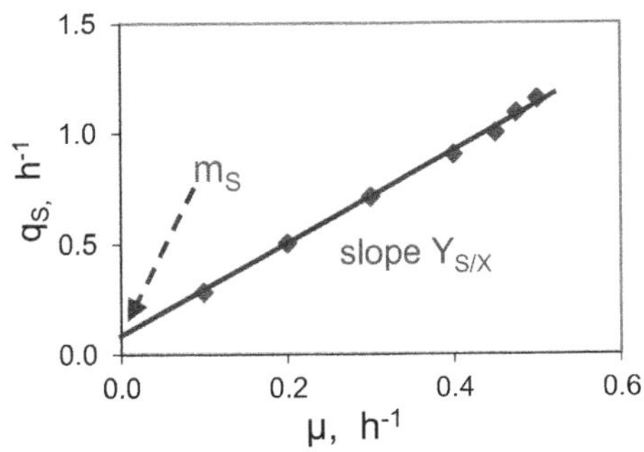

Figure 1.21. *Representation of the specific sugar consumption rate as a function of μ for the determination of the constants $Y_{X/S}$ and m_S*

1.3.3.5. *Multi-state kinetic models*

When cells undergo significant metabolic changes during fermentation, a relevant approach is to use kinetic models that include several physiological or metabolic states for the cells. In this type of model, cells pass through different phases: first, a latent state, then they gradually evolve toward a rapid growth phase, ultimately reaching a metabolite production state. These transitions between states can extend over periods of several hours.

Multi-state kinetic models incorporate separate kinetic laws for each physiological state, taking into account growth, substrate consumption and metabolite production for each state. In addition, they include laws that describe the transition rates of cells between these different physiological states (Figure 1.22).

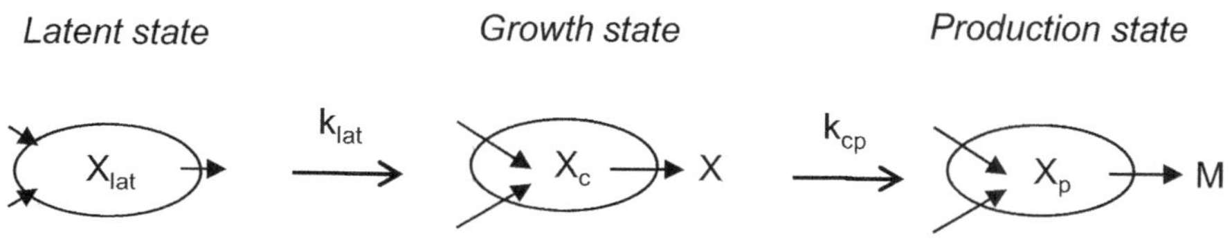

Figure 1.22. *Kinetic model with three physiological states of latency, growth and production for cells*

1.3.3.5.1. Kinetic modeling of the latent phase

A two-state physiological model can be used to simulate the initial latent phase of microbial growth. During this phase, the model assumes that the inoculated cells are in a latent state, with little or no metabolic and growth activity.

The gradual transition of latent cells X_{lat} to an active growth state X_c occurs at a transition volume rate r_{lat}''', expressed in g cell/L.h. The transition kinetics of latent cells follows a first-order law with a transition constant k_{lat}:

$$\text{rlat}''{}_{lat/X} = k_{lat}X_{lat} \qquad [1.54]$$

The value of the transition constant k_{lat} determines the duration of the latent phase: a higher value of k_{lat} results in a shorter latent phase, while a lower value leads to a longer latent phase.

1.3.3.5.2. Modeling of nongrowth metabolite production

Another application of the physiological state model is to model fermentations characterized by rapid initial cell growth, followed by a subsequent phase of metabolite excretion when growth is reduced or stopped.

In addition to the kinetic laws that are specific to each state, the model includes a law describing the rate of transition of cells from the growth state to the production state. Many factors can induce this metabolic transition. They include substrate limitations, such as those related to the supply of carbon, nitrogen, phosphorus or vitamins. In addition, changes in temperature or pH, the addition of inducers (for example, for protein synthesis) and effectors (such as surfactants) can play a role. By incorporating the causal factor into the relation describing the rate of transition, the physiological model enables the two successive phases of growth and metabolite production to be accurately simulated.

1.4. Oxygen consumption, solubility and transfer in fermenters

In aerated fermenters, oxygen transfer and consumption processes play a central role, as they often determine the maximum achievable concentrations of microbial cells and metabolites.

Oxygen consumption during aerobic fermentation is high, which can, in order of magnitude, be compared to the quantities of sugar used. However, oxygen has a very low solubility in fermentation media, requiring a continuous supply from the air and an efficient transfer from the air to the fermentation medium. Successful aerobic fermentation therefore depends on the ability of oxygen transfer, controlled by matter transfer, to meet the oxygen requirements of microbial cells (Doran 2013).

This section deals with the modeling of oxygen dynamics in microbial fermenters. It will successively address the rate of oxygen consumption by cells, the solubility of oxygen in fermentation media and the rate of oxygen transfer in fermenters.

1.4.1. *Oxygen consumption rates*

1.4.1.1. *Specific, volumetric and overall rates*

Oxygen consumption in a fermenter is characterized by three rate values. The first of these is the specific rate of oxygen consumption by microbial cells, q_{O2}, expressed in grams of oxygen consumed per gram of cells per hour. This rate is determined by the metabolic activities of the cells and can be quantified by the following kinetic law:

$$q_{O2} = Y_{O2/X}\mu + m_{O2} + Y_{O2/M}q_M \qquad [1.55]$$

The specific rate qO_2 includes a growth-related oxygen consumption term in which $Y_{O2/X}$ is the stoichiometric coefficient in grams of oxygen consumed per gram of cells produced. Common values for $Y_{O2/X}$ range from 0.6 to 0.8. The specific rate can also include a m_{O2} maintenance consumption term. In cases where metabolite production is not associated with cell growth, the law implies an additional term of oxygen consumption that is proportional to the metabolite production rate. This term is characterized by the stoichiometric coefficient $Y_{O2/M}$, which indicates the amount of oxygen consumed per gram of metabolite produced.

The second rate value used is the volumetric rate r_{O2}''' in grams of oxygen per liter per hour. The volumetric rate is the specific rate multiplied by the concentration of microbial cells. It represents the intensity of oxygen consumption in the fermenter.

The third value is the overall oxygen consumption rate r_{O2}, in grams of oxygen per hour, in the entire fermentation medium. The overall rate is the volumetric rate multiplied by the volume of the medium.

The three oxygen consumption rates q_{O2}, r_{O2}''' and r_{O2} evolve during microbial fermentation, in correlation with variations in the specific rates of cell growth and metabolite production.

1.4.1.2. *Determination of oxygen consumption rate using the steady-state balance*

1.4.1.2.1. Instantaneous oxygen balance of the fermenter

The first method for measuring the rate of oxygen consumption during fermentation is to use the instantaneous oxygen balance on the fermenter. As air is continuously added, the fermenter behaves as a continuous system for oxygen, even though it is operating in batch mode (Figure 1.23).

During fermentation, the overall rate of oxygen consumption r_{O2} is determined from the instantaneous oxygen balance:

$$Q_{O2,input} - r_{O2} = Q_{O2,output} + \frac{dm_{O2}}{dt} \quad [1.56]$$

where $Q_{O2,input}$ and $Q_{O2,output}$ are the oxygen input and output rates, and dm_{O2}/dt is the rate of change of oxygen mass in the fermenter.

Due to the very low solubility of oxygen, the fourth term of the internal rate of change is much lower than the other three terms of the balance relating to oxygen input, output and consumption rates. Consequently, this term can be ignored, allowing us to assume a steady state for oxygen in the fermenter.

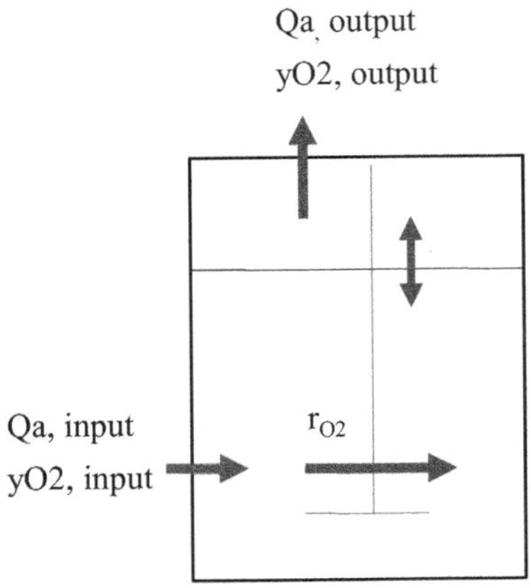

Figure 1.23. *Instantaneous oxygen balance terms for an aerated fermenter. For a color version of this figure, see www.iste.co.uk/ghoul/bioreactor.zip*

1.4.1.2.2. Oxygen input rate in the fermenter

Air is introduced into the fermenter at an input volume flow rate $Q_{a,input}$, which is measured in liters per hour. The air input molar flow rate is calculated by dividing the volume flow rate by the air molar volume $V_{m,input}$ under input conditions. At pressure P and temperature T, the air molar volume V_m is determined from the standard molar volume V_{m0} of 22.4 L at 0°C and 1 atm, using the ideal gas law:

$$V_m = V_{m0} \frac{T}{T_0} \frac{P_0}{P} \quad [1.57]$$

When the input air contains the mole fraction of oxygen $y_{O2,input}$, the oxygen input rate $M_{O2,input}$ is calculated by the following relation:

$$M_{O2,input} = 32 \frac{Q_{a,input}}{V_{m,input}} y_{O2,input} \quad [1.58]$$

1.4.1.2.3. Oxygen output rate of the fermenter

At the fermenter output, the air molar flow rate is the volumetric flow rate $Q_{a,output}$ divided by the air molar volume $V_{m,output}$ at output air temperature and pressure. This flow rate is generally different from the input molar flow rate due to the consumption of oxygen by microorganisms and the simultaneous production of carbon dioxide. The variation in air molar flow rate in the fermenter can be assessed from measurements of the molar fractions of oxygen and CO_2 in the input and output air using the instantaneous balance for nitrogen. Since nitrogen is neither consumed nor produced, the nitrogen molar flow rate at the input is equal to the nitrogen molar flow rate at the output, in other words:

$$\frac{Q_{a,input}}{V_{m,input}} y_{N2,input} = \frac{Q_{a,output}}{V_{m,output}} y_{N2,output} \qquad [1.59]$$

By expressing the molar fractions of nitrogen $y_{N2,input}$ and $y_{N2,output}$ in terms of the corresponding molar fractions of oxygen y_{O2} and carbon dioxide y_{CO2}, the nitrogen balance is written as follows:

$$\frac{Q_{a,input}}{V_{m,output}}\left(1 - y_{O2,input} - y_{CO2,input}\right) = \frac{Q_{a,output}}{V_{m,output}}\left(1 - y_{O2,output} - y_{CO2,output}\right) \qquad [1.60]$$

Thus, the air molar flow rate at the fermenter output is determined by the relation:

$$\frac{Q_{a,output}}{V_{m,output}} = \frac{Q_{a,input}}{V_{m,input}} \frac{\left(1 - y_{O2,input} - y_{CO2,input}\right)}{\left(1 - y_{O2,output} - y_{CO2,output}\right)} \qquad [1.61]$$

The variation in air molar flow rate in the fermenter can also be expressed in terms of the cell's respiratory quotient (RQ). The RQ is defined as the ratio between the specific molar rates of CO_2 production and oxygen consumption:

$$RQ = \frac{q_{CO2}}{q_{O2}} \qquad [1.62]$$

The air molar flow rate in the fermenter varies with the value of RQ. When RQ is greater than 1, the air flow rate increases, while it decreases when RQ is less than 1. To determine the air flow rate at the output, we can combine the three instantaneous balances for nitrogen, oxygen and carbon dioxide in the fermenter. This combination yields the following relation:

$$\frac{Q_{a,output}}{V_{m,output}} = \frac{Q_{a,input}}{V_{m,input}} \frac{1 + (RQ-1)\, y_{O2,input}}{1 + (RQ-1)\, y_{O2,output}} \qquad [1.63]$$

When the RQ is close to 1, the air molar flow rate at the output can be considered equal to the air molar flow rate at the input.

If the output air contains the oxygen mole fraction $y_{O2,output}$, the oxygen output rate $M_{O2,output}$ is calculated by the relation:

$$M_{O2,output} = 32\frac{Q_{a,output}}{V_{m,output}}y_{O2,output} \quad [1.64]$$

1.4.1.2.4. Oxygen consumption rate in the fermenter

Ignoring the internal mass variation term, the instantaneous oxygen balance in the fermenter is written as:

$$32\frac{Q_{a,input}}{V_{m,input}}\, y_{O2,input} - r_{O2} = 32\frac{Q_{a,output}}{V_{m,output}}\, y_{O2output} \quad [1.65]$$

As a result, the oxygen consumption rate r_{O2} in the whole fermenter is determined as the difference between the oxygen input and output rates:

$$r_{O2} = 32\frac{Q_{a,input}}{V_{m,input}}y_{O2,input} - 32\frac{Q_{a,output}}{V_{m,output}}\, y_{O2,output} \quad [1.66]$$

When the RQ is close to 1, the overall rate is simply calculated by the following relation:

$$r_{O2} = 32\frac{Q_{a,input}}{V_{m,input}}(y_{O2,input} - y_{O2,output}) \quad [1.67]$$

1.4.1.2.5. Example: determination of oxygen consumption rates during fermentation

During batch fermentation to produce a metabolite, oxygen consumption is monitored by measuring the mole fraction of oxygen in the output air of the fermenter. These data are shown in Figures 1.24(a) and (b).

Assuming a RQ close to 1, we determine the temporal variations in volumetric and specific oxygen consumption rates.

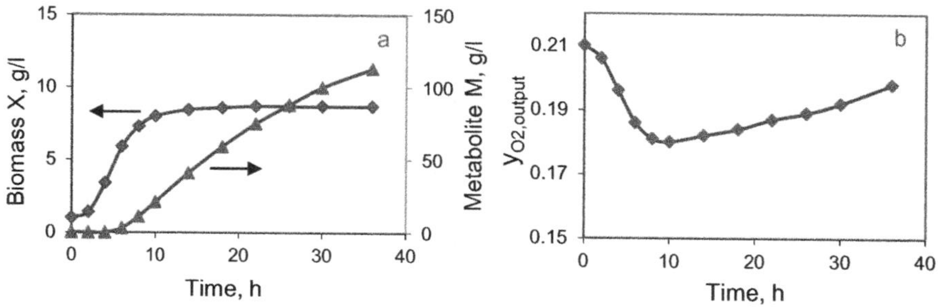

Figure 1.24. *(a) Variation in the concentrations of biomass X and metabolite M. (b) Variation in mole fraction of oxygen in the output air* $y_{O2,output}$

During fermentation, the oxygen molar fraction is constant at the input, around 0.21, while the output molar fraction, $y_{O2,output}$, decreases during the cell growth phase, before increasing during the metabolite production phase (Figure 1.24(b)).

The oxygen consumption rate r_{O2} in the fermenter is calculated as the difference between the oxygen input and output rates. The specific rate q_{O2}, determined by dividing the overall rate by the volume of the medium, increases during the initial phase of cell growth, reaches a maximum value and then gradually decreases during the metabolite production phase (Figure 1.24(c)).

The specific rate q_{O2}, which is calculated by dividing the volumetric rate by the biomass concentration, increases during the initial latent phase, then reaches a maximum value during the rapid cell growth phase, before decreasing during the metabolite production phase (Figure 1.24(d)).

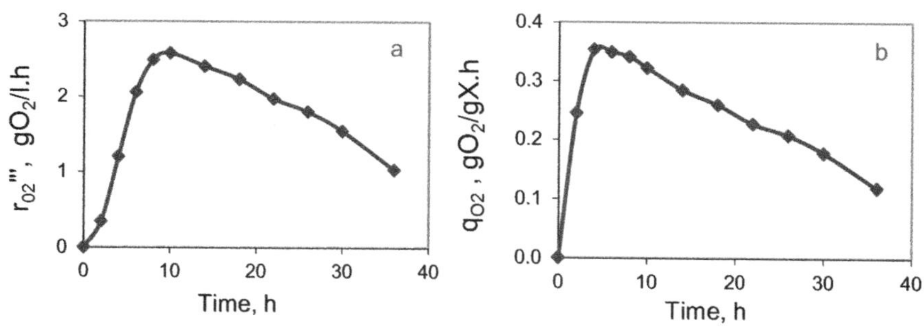

Figure 1.24. *(c) Variation of the volume rate of oxygen consumption* r_{O2}''' *over time. (d) Variation of the specific oxygen consumption rate* q_{O2} *over time*

1.4.1.3. *Measurement of oxygen consumption rate using the dynamic method*

1.4.1.3.1. Modeling the dynamic method

A second method for assessing the rate of oxygen consumption during fermentation is the dynamic method. In this approach, the air supply is momentarily cut off and the resulting decrease in oxygen concentration in the medium is measured using a dissolved oxygen probe.

The variation in oxygen concentration over time is modeled by writing the instantaneous oxygen balance in the liquid phase of the fermenter. When aeration is interrupted, no oxygen transfer occurs from the air to the liquid medium, and the instantaneous oxygen balance relation is as follows:

$$- r_{O2} = \frac{dm_{O21}}{dt} \qquad [1.68]$$

where m_{O21} is the mass of dissolved oxygen in the medium.

When dividing by the volume of the medium, the balance relation becomes:

$$-r'''_{O2} = \frac{dO_{21}}{dt} \qquad [1.69]$$

The oxygen consumption rate can be considered as constant during the brief period without aeration. Consequently, the dissolved oxygen concentration decreases linearly with time, and the oxygen consumption rate r_{O2}''' is determined by the slope of the straight line, representing the decrease in oxygen concentration as a function of time.

1.4.1.3.2. Influence of dissolved oxygen probe response time

When measuring oxygen concentration in the medium using a dissolved oxygen probe, the experimental variation in this concentration may result in a curve other than a straight line due to the probe's response time (Figure 1.25).

Oxygen probes used in fermenters generally feature an electrode, separated from the fermentation medium by a membrane. Dissolved oxygen in the medium diffuses through the membrane and the internal electrolyte solution to the electrode, where it reacts to produce a current that is proportional to the oxygen concentration in the medium. Given the time needed for oxygen diffusion, the probe exhibits a delay in its response to changes in dissolved oxygen levels. Commercially available steam sterilizable electrodes generally have response times ranging from a few dozen seconds to a minute.

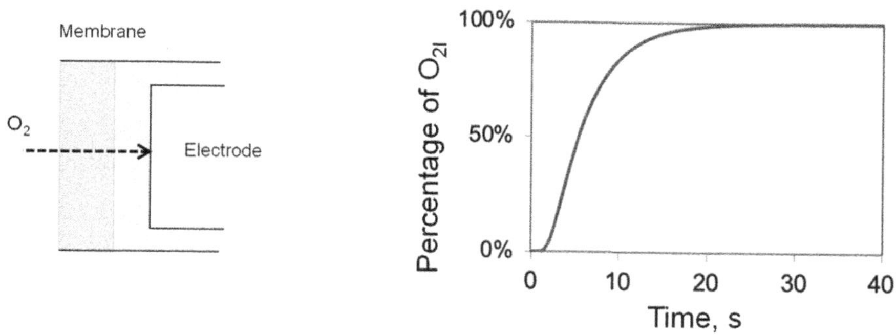

Figure 1.25. *Diagram of a dissolved oxygen probe and its response to dissolved oxygen transferred from a non-oxygenated solution to an oxygen-saturated solution*

This probe response time has no significant impact on oxygen measurements during fermentation processes, as variations in dissolved oxygen concentration generally occur over periods of several hours. However, when using the dynamic method, the decrease in oxygen concentration occurs more rapidly over a period of a few dozen seconds, which is comparable to the response time of the probe. Under such conditions, the decrease in oxygen concentration measured after aeration is stopped no longer follows a linear curve. Assessing the rate of oxygen consumption on the basis of the slope of a straight line is therefore erroneous, especially if the probe has a long response time (Figure 1.26).

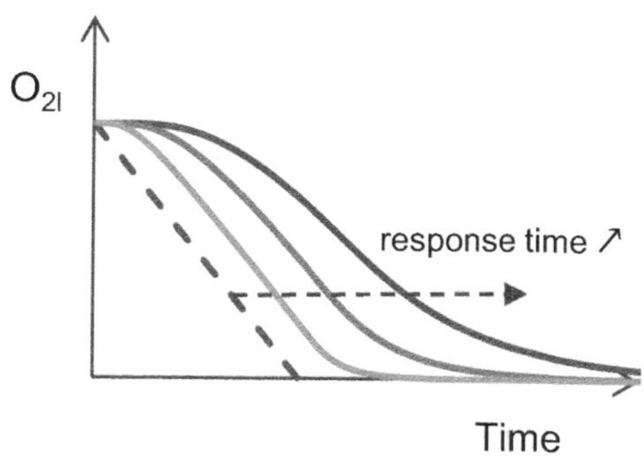

Figure 1.26. *Influence of probe response time on the variation in dissolved oxygen concentration, measured after stopping aeration in a fermenter*

To overcome this, various methods for processing experimental measurements exist based on modeling the dynamics of oxygen diffusion through the probe membrane. Through these approaches, it is possible to obtain a more accurate assessment of the oxygen consumption rate, taking into account the behavior of the probe during measurement.

1.4.2. *Oxygen solubility in fermentation media*

1.4.2.1. *Henry's law of solubility for oxygen in water*

Henry's law of solubility for oxygen in water states the proportionality between the oxygen content in air, expressed by its partial pressure P_{O2}, and the oxygen content in water, expressed by its molar fraction x_{O2}:

$$P_{O2} = H_{O2}\, x_{O2} \qquad [1.70]$$

where H_{O2} is Henry's law constant of oxygen in atmosphere when P_{O2} is in the atmosphere.

The partial pressure of oxygen is the product of the molar fraction of oxygen y_{O2} in air and the total pressure P:

$$P_{O2} = y_{O2} P \qquad [1.71]$$

1.4.2.2. *Henry's law with oxygen mass solubility (O_2*)*

Henry's law can be reformulated to assess oxygen solubility in grams per liter. This relation between the mole fraction x_{O2} and the solubility O_2* in gO_2/L can be derived as follows.

The x_{O2} mole fraction of oxygen in water is defined as:

$$x_{O2} = \frac{\text{moles } O_2}{\text{moles } O_2 + \text{moles } H_2O} \qquad [1.72]$$

As a solution, the number of oxygen molecules is much lower than the number of water molecules. Consequently, the oxygen mole fraction x_{O2} can be roughly expressed as:

$$x_{O2} = \frac{\text{moles } O_2}{\text{moles } H_2O} = \frac{\text{moles } O_2/L}{\text{moles } H_2O/L} \qquad [1.73]$$

The concentration of water is 55.5 mol/L (1,000 g/L divided by 18 g/mol). The molar solubility of oxygen in water O_2* in mol/L is therefore equal to:

$$O_2^* = 55.5\ x_{O2} \qquad [1.74]$$

and the O_2* mass solubility in grams of O_2 per liter:

$$O_2^* = 32 * 55.5\ x_{O2} \qquad [1.75]$$

Combining this relation with Henry's law, we obtain the modified Henry's law giving the O_2* mass solubility as a function of the partial pressure of oxygen P_{O2} in the air:

$$O_2^* = \frac{P_{O2}}{K_{O2}} \qquad [1.76]$$

In this relation, the solubility constant K_{O2} in atm/gO_2 is calculated from the Henry constant H_{O2} in atm, according to the following relation:

$$K_{O2} = \frac{H_{O2}}{32 * 55.5} \qquad [1.77]$$

1.4.2.3. *Influence of temperature on oxygen solubility*

The concentration of oxygen at equilibrium between air and water depends on temperature. Henry's law constant H_{O2} increases with temperature. Consequently, at a constant partial pressure of oxygen in the air, the solubility of oxygen in water decreases with temperature.

The values of Henry's law constant and the solubility of oxygen in water at different temperatures are shown in Table 1.1, for an air pressure of 1 atm.

Temperature (°C)	H atm	O_2* mg/L
10	32,000	11.7
20	40,000	9.3
30	47,000	7.9
40	53,000	7

Table 1.1. *Oxygen solubility according to Henry's law*

1.4.2.4. *Oxygen solubility in fermentation media*

The presence of solutes such as salts, acids and sugars reduces oxygen solubility in fermentation media. Empirical correlations exist for correcting oxygen solubility values in water, while considering the effects of concentrations of added ionic and nonionic compounds. For example (Doran 2013):

$$\log_{10}\left(\frac{O^*_{2,water}}{O^*_2}\right) = 0.5 \sum_i H_i z_i^2 C_i + \sum_j K_j C_j \qquad [1.78]$$

where $O_2{}^*_{water}$ is the solubility of oxygen in water, $O_2{}^*$ is the solubility of oxygen in the medium, H_i is the constant for ionic compound i, z_i is the charge of ionic compound i, C_i is the concentration of ionic compound (mole/m^3), K_j is the constant for nonionic compound j and C_j is the concentration of nonionic compound (mole/m^3).

The reductions in oxygen solubility calculated using this relation for a glucose solution are illustrated in Figure 1.27. It can be seen that oxygen solubility is reduced by up to 50% at a concentration of 400 g/L sugar. Generally speaking, in a fermentation medium, oxygen solubility is between 10% and 25% lower than in water, due to the effects of solutes.

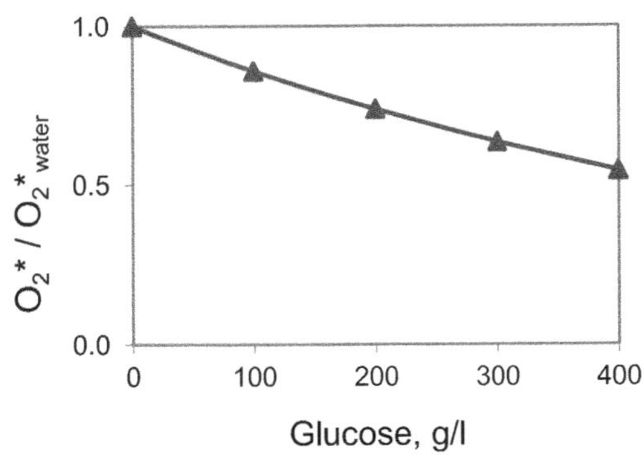

Figure 1.27. *Influence of glucose concentration on oxygen solubility $O_2{}^*$ normalized to water solubility $O_2{}^*_{water}$*

1.4.2.5. *Average oxygen solubility in a fermenter*

When air is introduced at the bottom of an agitated fermenter, the partial pressure of oxygen varies between the fermenter input and output. To take these variations into account, it is preferable to use an average partial pressure of oxygen in order to calculate the average solubility of oxygen in the fermenter.

The P_{input} air pressure at the bottom of the fermenter is higher than the P_{output} pressure at the top of the fermenter. If the height of the medium in the fermenter is h_m, the P_{input} pressure is determined by the following relationship:

$$P_{input} = P_{output} + \rho g\, h_m \qquad [1.79]$$

where ρ is the density of the medium and g is the gravitational acceleration constant.

In the input air, at the oxygen mole fraction $y_{O2,input}$, the partial pressure of oxygen is:

$$P_{O2,input} = y_{O2,input}\, P_{input} \qquad [1.80]$$

At the fermenter output, the oxygen molar fraction $y_{O2,output}$ is lower than at the input, and the partial pressure of oxygen is:

$$P_{O2,output} = y_{O2,output}\, P_{output} \qquad [1.81]$$

For the entire fermenter, the average partial pressure of oxygen $P_{O2,m}$ is introduced, which is calculated according to the following formula:

$$P_{O2,m} = \frac{P_{O2,input} + P_{O2,output}}{2} \qquad [1.82]$$

Consequently, the average solubility of oxygen $O_2{}^*{}_m$ is determined using the solubility law:

$$O^*_{2,m} = \frac{P_{O2,m}}{K_{O2}} \qquad [1.83]$$

1.4.3. *Oxygen transfer rate in a fermenter*

1.4.3.1. *Stages of oxygen transfer between the air and liquid medium*

Inside a fermenter, oxygen is continuously transferred from the gas phase to the liquid medium. There are three stages involved in oxygen transfer: (i) transfer by diffusion and convection from the air to the air–liquid interface; (ii) the dissolution process itself, during which oxygen molecules in the air are transferred to the first liquid layer at the interface; (iii) transfer of oxygen by diffusion and convection from the interface to the solution (Figure 1.28(a)).

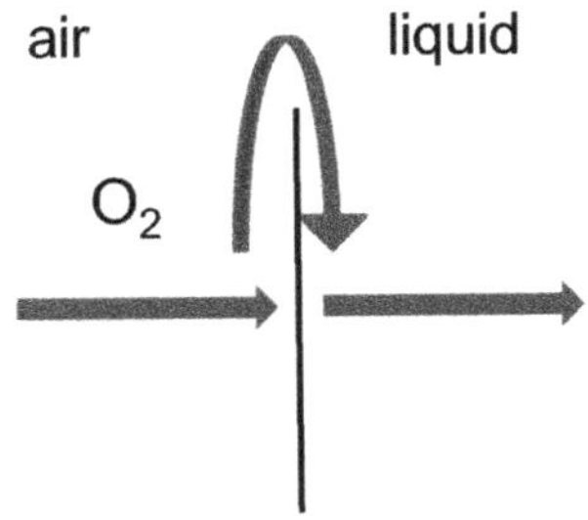

Figure 1.28. *(a) Stages of oxygen transfer between the air and fermentation medium. For a color version of this figure, see www.iste.co.uk/ghoul/bioreactor.zip*

The three stages present different resistances to oxygen transfer. The dissolution process itself, which involves the transfer of oxygen molecules from the gas interface to the first liquid layer, is very easy. Similarly, the transfer of oxygen into the gas phase is straightforward. However, the most difficult stage, which limits the overall rate of transfer, is the passage of oxygen from the liquid phase of the interface to the solution.

1.4.3.2. *Oxygen concentration profile at the air–liquid interface*

Given that the resistance to oxygen transfer in the gas phase is very low, there is no oxygen gradient in the air up to the interface. The partial pressure of oxygen P_{O2} is identical throughout the gas phase (Figure 1.28(b)).

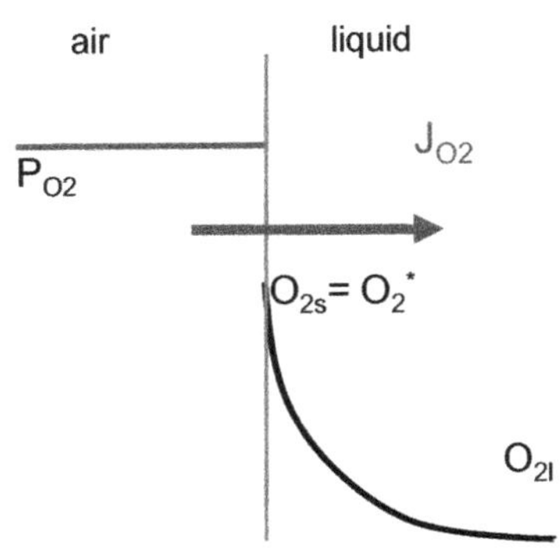

Figure 1.28. *(b) Oxygen concentration profile at the air-liquid interface. For a color version of this figure, see www.iste.co.uk/ghoul/bioreactor.zip*

The resistance of the oxygen dissolution stage is also negligible. Under these conditions, a maximum number of oxygen molecules are transferred to the first interfacial liquid layer, until the solubility of oxygen in water is reached. The dissolved oxygen concentration at the surface O_{2s} is therefore equal to the solubility O_2* that is in equilibrium with the partial pressure of oxygen P_{O2} in the air.

In the liquid phase, which represents the limiting stage, an oxygen concentration gradient is established between the O_2* solubility at the interface and the O_{2l} oxygen concentration in the liquid medium.

1.4.3.3. *Kinetic law of oxygen transfer*

The rate of oxygen transfer throughout the liquid medium, denoted as J_{O2} in kgO_2/s or gO_2/h, is expressed by the kinetic law of mass transfer by diffusion and convection at the air–liquid interface:

$$J_{O2} = Sk_l\,(O_2^* - O_{2l}) \qquad [1.84]$$

where S (m^2) is the transfer surface between the air and the liquid solution, and k_l (m/s) is the oxygen transfer coefficient in the liquid solution. The transfer surface S encompasses both the area in contact with the air above the medium and the surface of circulating air bubbles within the medium. The difference in oxygen concentration $O_2* - O_{2l}$ is the driving force for oxygen transfer.

By introducing the interface surface area a (m^2/m^3) between the air and the liquid medium:

$$a = \frac{S}{V} \qquad [1.85]$$

The oxygen transfer rate J_{O2} becomes:

$$J_{O2} = k_l a(O_2^* - O_{2l})V \qquad [1.86]$$

The coefficient k_la (s^{-1} or h^{-1}) is the volumetric coefficient of oxygen transfer. It characterizes the oxygen transfer capacity of the aeration system.

1.4.3.4. *Values of the oxygen transfer coefficient kla in fermenters*

In a fermenter, the oxygen transfer coefficient k_la is influenced by various factors. Among these, the agitation of the medium plays a key role, in particular the geometry and rate of rotation of the agitators. Aeration conditions, determined by the geometry of the air diffuser (perforated tube or microporous membrane) and the air flow rate, as well as the viscosity and surface tension of the medium, are also decisive factors (van't Riet 1983).

Under stable conditions of agitation and aeration, k_la can generally be maintained at a relatively constant level during fermentation. However, it can decrease considerably if the viscosity of the medium increases, for example, due to the excretion of polysaccharides or changes in the morphology of filamentous bacteria or fungi. Similarly, the addition of anti-foaming agents can reduce k_la by modifying the surface tension of the medium.

In microbial culture systems, volumetric oxygen transfer coefficients k_la generally range from 1 to 1,000 h^{-1}. The lowest values are found in surface aeration systems, intermediate values in bubble columns, and the highest values in mechanically agitated fermenters.

In mechanically agitated fermenters, the oxygen transfer coefficient can be estimated from correlations expressing k_la as a function of agitation power density P_{ag}/V and air surface velocity u_g. Surface velocity is defined as the ratio of air volume flow Q_a to the cross-sectional area of the fermenter Ω:

$$u_g = \frac{Q_a}{\Omega} \quad [1.87]$$

An often-used correlation is the following form:

$$k_la = A\left(P_{ag}/V\right)^m u_g^n \quad [1.88]$$

where the constants A, m and n depend on fermenter size, agitation geometry, medium viscosity and surface tension. Figure 1.29 shows the k_la values predicted by this correlation, for an industrial-sized fermenter, according to agitation power density and air surface velocity.

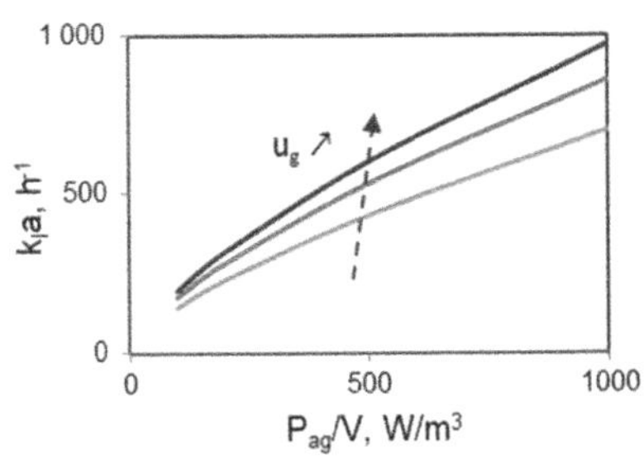

Figure 1.29. *Influence of agitation power density P_{ag}/V and surface air velocity u_g on oxygen transfer volume coefficient k_la*

1.4.3.5. *Measuring k_la using the steady-state method*

This k_la assessment method uses the oxygen balance of the liquid phase inside the fermenter. Since oxygen is continuously transferred from the gas phase to the liquid phase at a rate of J_{O2}, the instantaneous oxygen balance of the liquid phase is:

$$J_{O2} - r_{O2} = \frac{dm_{O2l}}{dt} \quad [1.89]$$

where m_{O2l} is the mass of oxygen dissolved in the liquid medium. Due to the low solubility of oxygen, the rate of change term is negligible, and the oxygen balance becomes:

$$J_{O2} = r_{O2} \quad [1.90]$$

This relationship means that at every instant during fermentation, a steady state is established, where the oxygen transfer rate J_{O2} is equal to the oxygen consumption rate r_{O2}. Introducing the kinetic law of oxygen transfer, the oxygen balance becomes:

$$k_la\left(O^*_{2,m} - O_{2l}\right)V = r'''_{O2}V \quad [1.91]$$

The oxygen transfer coefficient k_la is calculated by the following equation:

$$k_la = \frac{r'''_{O2}}{(O^*_{2,m} - O_{2l})} \quad [1.92]$$

Measurement of the k_la transfer coefficient during fermentation requires simultaneous determination of the volumetric rate of oxygen consumption, using the oxygen balance of the fermenter, as well as the concentration of dissolved oxygen, which is measured using an oxygen probe.

The steady–state oxygen balance method is the most reliable procedure for estimating k_la. One of its major advantages is that it can be applied to fermenters in normal operation. However, it is essential to note that this method is highly dependent on the accuracy of air composition measurements at the fermenter input and output, as well as flow rate, pressure and temperature measurements.

1.4.3.6. *Determination of k_la using the dynamic method*

The second method for measuring k_la is the dynamic method. During fermentation, the air supply is interrupted to allow the dissolved oxygen concentration to decrease. The air supply is then reestablished and the increase in dissolved oxygen concentration is measured with the oxygen probe. After a relatively short period, the dissolved oxygen level returns to its initial value before aeration was interrupted.

The temporal variation in dissolved oxygen concentration, before and after aeration is interrupted, is modeled by the instantaneous oxygen balance of the liquid medium (Figure 1.30(a)).

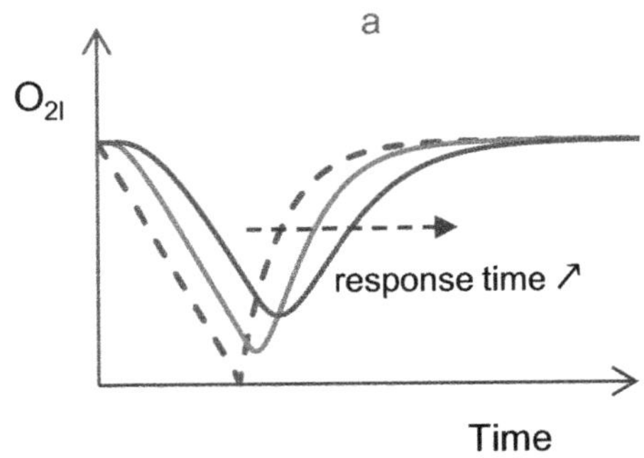

Figure 1.30. *(a) Influence of probe response time on the variation in dissolved oxygen concentration, measured after aeration of a fermenter has been stopped and then restarted*

Before aeration stops, the rate of change of dissolved oxygen mass is negligible. As a result, the steady-state dissolved oxygen concentration $O_{21,ss}$ is determined by the equality of the oxygen transfer and consumption rates:

$$k_l a \left(O^*_{2,m} - O_{21,ss}\right) = r'''_{O2} \quad [1.93]$$

During the dynamic experiment, when aeration is restored, the rate of change of dissolved oxygen concentration is much higher and the instantaneous oxygen balance in the liquid phase is:

$$k_l a\left(O^*_{2,m} - O_{21}\right) - r'''_{O2} = \frac{dO_{21}}{dt} \quad [1.94]$$

Combining both steady-state and dynamic oxygen balances, the differential equation for dissolved oxygen concentration becomes:

$$k_l a\left(O_{21,ss} - O_{21}\right) = \frac{dO_{21}}{dt} \quad [1.95]$$

Or:

$$k_l a dt = \frac{dO_{2l}}{(O_{2l,ss} - O_{2l})} \qquad [1.96]$$

This differential equation is integrated between times t_0 and t to obtain the following relationship:

$$k_l a(t - t_0) = \ln\left(\frac{(O_{2l,ss} - O_{2l0})}{(O_{2l,ss} - O_{2l,})}\right) \qquad [1.97]$$

According to this relationship, $k_l a$ can be defined graphically by plotting the second logarithmic term of the equation as a function of time. The $k_l a$ corresponds to the slope of the straight line obtained in this representation.

As previously mentioned when determining the oxygen consumption rate using the dynamic method, it is crucial to consider the probe response time, which can greatly influence the measurement of the variation in dissolved oxygen concentration (Pouliot et al. 2000). Indeed, a high probe response time will result in a significant discrepancy between the theoretical variation and the measured variation (Figure 1.30(b)). This can lead to significant errors in the assessment of the $k_l a$ transfer coefficient by the dynamic method. To get a more accurate assessment of $k_l a$, it is possible to use methods for processing experimental measurements based on the modeling of the dynamics of oxygen diffusion in the probe.

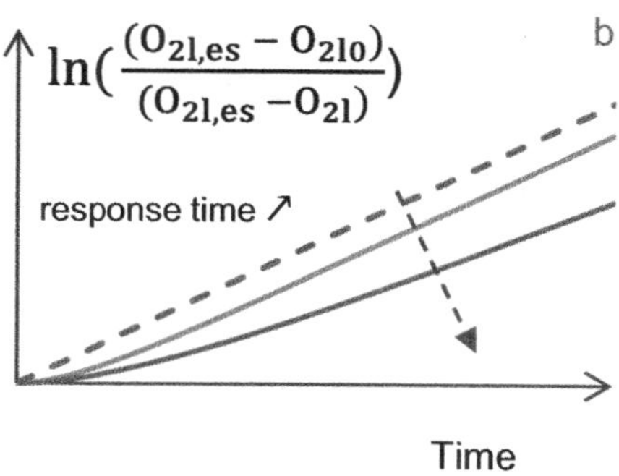

Figure 1.30. *(b) Influence of probe response time on the representation used to determine the oxygen transfer coefficient $k_l a$ by the dynamic method*

1.4.4. *Variation in dissolved oxygen concentration*

1.4.4.1. *Equal oxygen transfer and consumption rates*

The variation in dissolved oxygen concentration during aerobic fermentation is governed by three main factors: the rate of oxygen transfer from the air to the medium, the solubilization of oxygen in the medium and the rate of oxygen consumption by the cells. This variation is assessed more accurately from the instantaneous steady-state oxygen balance in the liquid medium.

$$J_{O2} = r_{O2} \qquad [1.98]$$

According to this relation, which is key to the kinetics of aerobic fermentations, the rate of oxygen transfer must, at each instant, be equal to the rate of oxygen consumption by the cells.

However, the question is: how is this equality between oxygen transfer and consumption rates established instantaneously, when the oxygen consumption rate changes during fermentation?

When the oxygen transfer rate is expressed by its kinetic law, the equality becomes:

$$k_l a\left(O_{2,m}^{*} - O_{2l}\right) = r_{O2}''' \qquad [1.99]$$

According to this relation, it is the oxygen transfer driving force, in other words, the difference between the average oxygen solubility $O_2{}^*{}_m$ and the dissolved oxygen concentration O_{2l}, that is adjusted to meet the equality of oxygen transfer and consumption rates. For example, when the volumetric rate of oxygen consumption is increased, the J_{O2} transfer rate is also increased by decreasing the dissolved oxygen concentration (Figure 1.31(a)).

The variation in dissolved oxygen concentration is thus calculated from the temporal variation in the volumetric rate of oxygen consumption, according to the following relation (Figure 1.31(b)):

$$O_{2l} = O_{2,m}^{*} - \frac{r_{O2}'''}{k_l a} \qquad [1.100]$$

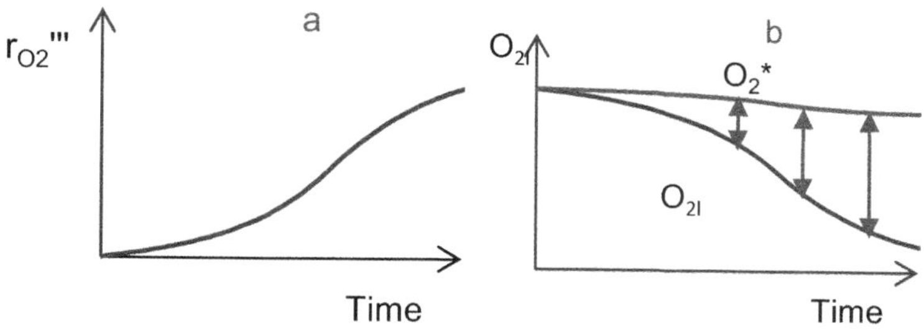

Figure 1.31. *(a) Influence of probe response time on the variation in dissolved oxygen concentration, measured after aeration of a fermenter has been stopped and then restarted. (b) Influence of probe response time on the representation used to determine the oxygen transfer coefficient kla using the dynamic method. For a color version of this figure, see www.iste.co.uk/ghoul/bioreactor.zip*

1.4.4.2. *Variation in dissolved oxygen concentration during metabolite fermentation*

During a typical metabolite fermentation, the volumetric rate of oxygen consumption initially increases during the cell growth phase, and then decreases during the metabolite production phase (Figure 1.32(a)).

The average oxygen solubility in the fermenter is not constant but varies slightly due to fluctuations in the partial pressure of oxygen in the output air, as well as variations in substrate and metabolite concentrations (Figure 1.32(b)).

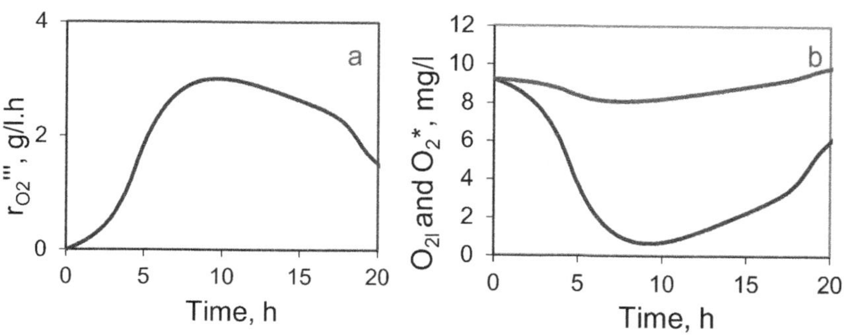

Figure 1.32. *(a) Variation of the volumetric rate of oxygen consumption r_{O2}''' over time. (b) Variation in the difference between oxygen solubility O_2^* and dissolved oxygen concentration O_{2l}*

In order to maintain equal oxygen transfer and consumption rates, dissolved oxygen concentration decreases during the rapid cell growth phase, reaching a minimum level corresponding to the peak oxygen consumption rate, and then increases again during metabolite production.

1.5. CO_2 production, solubility and transfer in fermenters

Carbon dioxide is a metabolite produced during both aerobic and anaerobic fermentation. In a fermenter, CO_2 undergoes various transfer and reaction stages. Initially, the cells excrete CO_2 into the fermentation medium. Some of this is hydrated to form carbonic acid, which then dissociates into bicarbonate and carbonate ions. Most of the CO_2 produced desorbs from the medium and passes into the gas phase, where it is removed from the fermenter (Dixon and Kell 1989a).

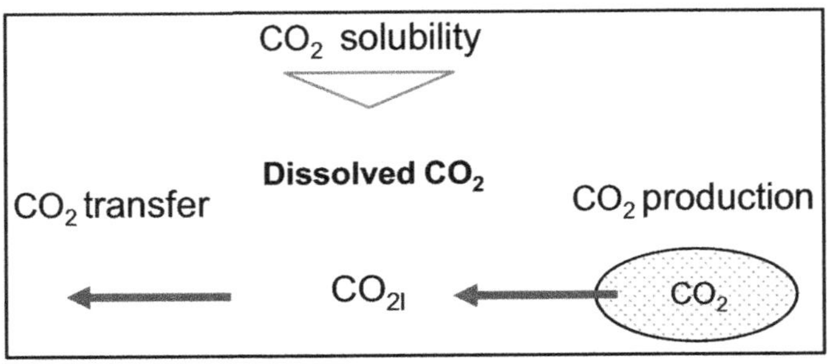

Figure 1.33. *Factors influencing the concentration of dissolved CO_2 in a fermentation medium*

Microbial CO_2 production can influence fermentation kinetics in two ways (Jones and Greenfield 1982; Dixon and Kell 1989b). First, it contributes to the acidification of the medium, and second, dissolved CO_2 can inhibit cell growth and metabolism. This inhibition becomes significant when dissolved CO_2 reaches relatively high levels in the medium. During fermentation, the level of dissolved CO_2 depends on three factors: microbial production, solubilization of CO_2 in the medium and transfer of CO_2 to the gas phase (Figure 1.33).

This next section deals with the modeling of CO_2 dynamics in microbial fermenters, taking into account the rate of CO_2 production, the solubility of CO_2 in the media and the rate of CO_2 transfer. It also describes the instantaneous CO_2 balances that govern the variation of CO_2 concentration inside the fermenter.

1.5.1. CO_2 production rate

1.5.1.1. *Kinetic law of CO_2 production*

Carbon dioxide production by microbial cells is characterized by the specific rate of production q_{CO2}, measured in grams of CO_2 produced per gram of cells per hour.

In aerobic cells, q_{CO2} is directly related to the specific rate of oxygen consumption:

$$q_{co2} = Y_{co2/O2} q_{o2} \quad [1.101]$$

where $Y_{CO2/O2}$ is the stoichiometric coefficient expressed in grams of CO_2 produced per gram of O_2 consumed. This mass coefficient can be estimated from the molar RQ by the following relation:

$$Y_{co2/o2} = 1.4RQ \quad [1.102]$$

The value of the RQ depends on the microbial strain, the substrates consumed and the physiological state of the cell and can vary during fermentation. For example, for the yeast *Saccharomyces cerevisiae*, RQ is close to 1 when the cell is developing during respiratory metabolism without ethanol excretion, but it is greater than 1 when it is in the fermentative state excreting ethanol.

In the case of anaerobic fermentations, the specific rate q_{CO2} is related to the specific rate q_S of carbonaceous substrate consumption:

$$q_{CO2} = Y_{co2/S}\, q_S \quad [1.103]$$

Here, $Y_{CO2/S}$ is the stoichiometric coefficient expressed in grams of CO_2 produced per gram of substrate consumed. For example, for anaerobic growth of *Saccharomyces cerevisiae* yeast, $Y_{CO2/S}$ is equal to 0.47 g of CO_2 per gram of glucose.

It is important to note that the specific rate of CO_2 production varies during fermentation, depending on changes in the specific rates of cell growth and metabolite production, for both aerobic and anaerobic fermentations.

1.5.1.2. *Determination of the CO_2 production rate using the steady-state balance*

As with oxygen, the rate of CO_2 production can be determined during fermentation using the steady-state balance method.

During fermentation, the overall rate of CO_2 production, noted as r_{CO2} and expressed in gCO_2/h, is determined from the instantaneous CO_2 balance in the entire fermenter:

$$M_{CO2,input} + r_{CO2} = M_{C2,output} + \frac{dm_{CO2}}{dt} \qquad [1.104]$$

In this balance, $M_{CO2,input}$ and $M_{CO2,output}$ are the input and output rates of CO_2 in the input and output air flows, and dm_{CO2}/dt is the rate of change of CO_2 mass in the fermenter.

Given that the solubility of CO_2 in the fermentation medium is very low, the rate of change of CO_2 mass in the fermenter is negligible compared to the rates of CO_2 input, output and production. Consequently, a quasi-steady state can be assumed for CO_2 in the fermenter.

Expressing the input and output rate terms according to the air molar flow rate and the corresponding molar fractions of CO_2, the balance is written as follows:

$$44\frac{Q_{a,input}}{V_{m,input}} y_{CO2,input} + r_{CO2} = 44\frac{Q_{a,output}}{V_{m,output}} y_{CO2,output} \qquad [1.105]$$

Consequently, the CO_2 production rate in the fermenter is determined as the difference between the input and output rates:

$$r_{CO2} = 44\frac{Q_{a,output}}{V_{m,output}} y_{CO2,output} - 44\frac{Q_{a,input}}{V_{m,input}} y_{CO2,input} \qquad [1.106]$$

If the RQ is close to 1, the CO_2 production rate can be simply calculated by the following relation:

$$r_{CO2} = 44\frac{Q_{a,input}}{V_{m,input}}\left(y_{CO2,output} - y_{CO2,input}\right) \qquad [1.107]$$

1.5.1.3. *Example: CO_2 production rates during metabolite fermentation*

During aerobic batch fermentation to produce a metabolite unrelated to cell growth, the mole fraction $y_{CO2,output}$ is measured in the output air (Figures 1.34(a) and (b)). Assuming an RQ close to 1, determine the temporal variations in volumetric and specific CO_2 production rates.

At the start of fermentation, during the microbial growth phase, the molar fraction $y_{CO2,output}$ increases. However, as metabolite production intensifies, this fraction decreases.

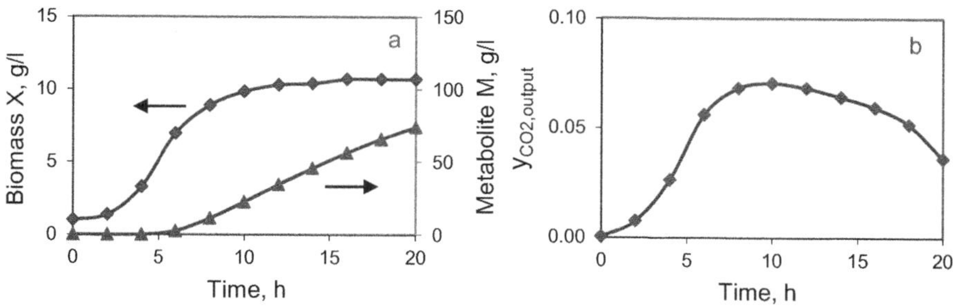

Figure 1.34. *(a) Variation in biomass X and metabolite M concentrations over time. (b) Variation in mole fraction of CO_2 in output air $y_{CO2,\ output}$*

The production rate r_{CO2} in the fermenter is calculated as the difference between the input and output rates of CO_2, and the volumetric rate r_{CO2}''' by dividing r_{CO2} by the volume V of the medium. The results show that the r_{CO2}''' volumetric rate increases at the start of the cell growth phase, reaches a maximum value and then gradually decreases during metabolite production (Figure 1.34(c)).

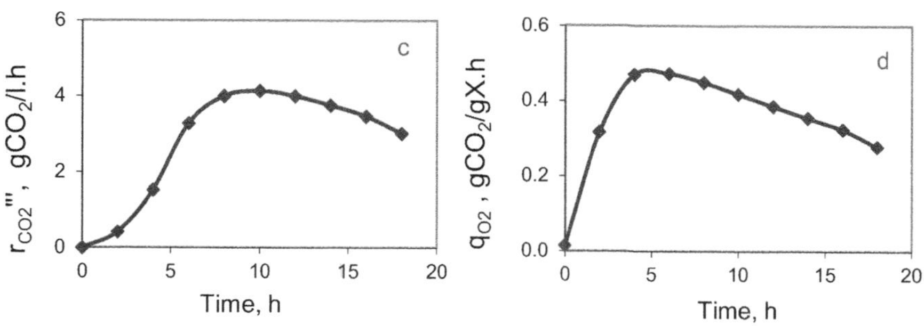

Figure 1.34. *(c) Variation of the volumetric rate r_{CO2}''' of CO_2 production over time. (d) Variation of the specific rate q_{CO2} of CO_2 production over time*

The specific production rate q_{CO2} is calculated by dividing the volumetric rate by the biomass concentration. It increases during the initial latent phase, reaches a maximum value during the rapid cell growth phase and then decreases during metabolite production (Figure 1.34(d)).

1.5.2. *Solubility and dissociation equilibria of CO_2*

1.5.2.1. *Henry's law for the solubility of CO_2 in water*

Henry's law establishes the proportional relationship between the partial pressure of P_{CO2} of CO_2 in air and its molar fraction x_{CO2} in water.

$$P_{CO2} = H_{CO2}x_{CO2} \quad [1.108]$$

The constant of proportionality H_{CO2} is Henry's constant in atm.

Henry's law can be modified to directly calculate CO_2* mass solubility in grams of CO_2 per liter, as a function of P_{CO2} partial pressure:

$$CO_2^* = \frac{P_{CO2}}{K_{CO2}} \quad [1.109]$$

The constant K_{CO2} is determined from Henry's constant H_{CO2} by the relation:

$$K_{CO2} = \frac{H_{CO2}}{44 * 55.5} \quad [1.110]$$

At 30°C, the K_{CO2} constant is equal to 0.68 atm/gCO_2.L.

1.5.2.2. *Hydration and dissociation of CO_2 in solution*

When carbon dioxide dissolves in water, it undergoes a reversible hydration reaction to form carbonic acid H_2CO_3 according to the following equation:

$$CO_2^* + H_2O \leftrightarrow H_2CO_3 \quad [1.111]$$

The hydration equilibrium constant, K_{H2CO3}, is defined as:

$$K_{H2CO3} = \frac{H_2CO_3}{CO_2} \quad [1.112]$$

It is equal to 1.7×10^{-3}. This indicates that the equilibrium strongly favors dissolved CO_2 over carbonic acid.

Carbonic acid is a weak acid that reversibly dissociates into bicarbonate HCO^{3-} and carbonate CO^{3-}. Let us first look at the dissociation of carbonic acid into bicarbonate:

$$H_2CO_3 \leftrightarrow HCO_3^- + H^+ \quad [1.113]$$

The equilibrium constant K_{a1} for this dissociation of carbonic acid is defined as:

$$K_{a1} = \frac{[HCO_3^-]\,[H^+]}{[H_2CO_3]} \qquad [1.114]$$

Its value is close to 2.5×10^{-4}. At pH 7, the ratio of bicarbonate to carbonic acid is:

$$\frac{[HCO_3^-]}{[H_2CO_3]} = \frac{K_{a1}}{[H^+]} = \frac{2.5\ 10^{-4}}{10^{-7}} = 2{,}500 \qquad [1.115]$$

As a result, the bicarbonate concentration is much higher than the carbonic acid concentration.

For the second dissociation of bicarbonate HCO^{3-} into carbonate CO^{3-}, the reaction is as follows:

$$HCO_3^- \leftrightarrow O_3^{2-} + H^+ \qquad [1.116]$$

The equilibrium constant K_{a2} for this dissociation is defined as:

$$K_{a2} == \frac{[CO_3^{2-}]\,[H^+]}{[HCO_3^-]} \qquad [1.117]$$

It is equal to 4.7×10^{-11}. Under these conditions, at pH 7, the ratio of carbonate to bicarbonate is equal to:

$$\frac{[CO_3^{2-}]}{[HCO_3^-]} = \frac{K_{a2}}{[H^+]} = \frac{4.7 \times 10^{-11}}{10^{-7}} = 4.7 \times 10^{-4} \qquad [1.118]$$

The carbonate concentration is therefore negligible when compared to the bicarbonate concentration.

1.5.2.3. *Average solubility of CO_2 in a fermenter*

In the same way as for oxygen solubility, an average CO_2 solubility is set in the fermenter.

For the entire fermenter, the average partial pressure of CO_2, noted as $P_{CO2,m}$, is calculated as follows:

$$P_{CO2,m} = \frac{y_{CO2,input}\ P_{input} + y_{CO2,output}\ P_{output}}{2} \qquad [1.119]$$

Under these conditions, the average solubility of CO_2, noted as $CO_2{}^*{}_m$ and expressed in gCO_2/L, is calculated by Henry's modified law:

$$CO^*_{2,m} = \frac{P_{CO2,m}}{K_{CO2}} \qquad [1.120]$$

1.5.3. *CO_2 transfer rate*

1.5.3.1. *Limiting stage and concentration gradients for CO_2 transfer*

As with oxygen, the transfer of carbon dioxide from the liquid medium to the gas phase involves three distinct stages: (i) transfer of CO_2 by diffusion and convection in the liquid medium; (ii) desorption of CO_2 at the interface; and (iii) transfer of CO_2 by diffusion and convection in the air (Figure 1.35).

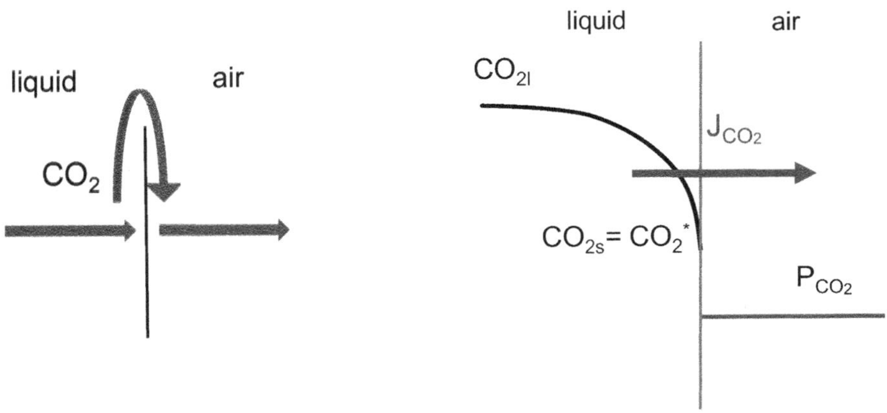

Figure 1.35. *Stages of CO_2 transfer between fermentation medium and air, and profile of dissolved CO_2 concentration and partial pressure of CO_2 between the medium and the air. For a color version of this figure, see www.iste.co.uk/ghoul/bioreactor.zip*

The last two stages, which are the desorption of CO_2 at the interface and its transfer to the air, are relatively easy and are characterized by low resistance. However, the first stage, transferring CO_2 into the liquid phase, is the most difficult stage and presents the greatest resistance.

As a result, there is no CO_2 gradient in the air between the interface and the gas phase. The partial pressure of CO_2, P_{CO2}, is uniform throughout the gas. The first liquid layer at the interface is saturated with CO_2, and the CO_2 concentration at the surface, CO_{2s}, is equal to the CO_2* solubility in equilibrium with the partial pressure P_{CO2} in the air. In the liquid phase, on the other hand, there is a gradient between the concentration of CO_2 that is dissolved in the medium, CO_{2l}, and the CO_2* solubility at the interface.

1.5.3.2. *Kinetic law for CO2 transfer rate between medium and air*

The rate of carbon dioxide transfer (J_{CO2}), expressed in kg CO_2/s or g CO_2/h, between the liquid medium and the air, is determined by the kinetic law of mass transfer by diffusion and convection:

$$J_{CO2} = Sk_l(CO_{2l} - CO_2^*) \qquad [1.121]$$

This rate is a function of the transfer surface S (m^2), between the air and the liquid solution, and of the transfer coefficient k_l (m/s) of the CO_2 in the liquid solution.

If we introduce the interfacial surface area a (m^2/m^3) between the air and the solution, the oxygen transfer rate J_{CO2} is then expressed by the following kinetic law:

$$J_{CO2} = kla(CO_{2l} - CO_2^*)V \qquad [1.22]$$

In this relation, k_la (s^{-1} or h^{-1}) is the volume transfer coefficient of CO_2 between the air and the liquid solution. The k_la coefficient for CO_2 can be considered approximately equal to the k_la coefficient for oxygen, since the molecular weights of oxygen and CO_2 are close.

1.5.4. *Variation in dissolved CO2 concentration during aerobic fermentation*

1.5.4.1. *Variation of CO2 solubility in the medium*

During aerobic fermentation, the solubility of CO_2 varies with the mole fraction of CO_2 in the output air or the rate of CO_2 production.

In a typical fermentation where the volumetric rate of CO_2 production increases during the growth phase and gradually decreases during the metabolite production phase, CO_2 solubility shows a similar variation: it increases during rapid cell growth and decreases during metabolite production.

1.5.4.2. *Variation in dissolved CO_2 concentration*

The temporal variation of the dissolved CO_2 concentration is assessed from the instantaneous CO_2 balance in the liquid medium:

$$r_{CO2} = J_{CO2} + \frac{dm_{CO2l}}{dt} \qquad [1.23]$$

Due to the low solubility of carbon dioxide, the rate of change of CO_2 mass is negligible, simplifying the CO_2 balance as follows:

$$r_{CO2} = J_{CO2} \qquad [1.24]$$

Consequently, at every instant during fermentation, a steady state is established in the liquid medium, and the production rate r_{CO2} is equal to the transfer rate J_{CO2}.

When the transfer rate J_{CO2} is expressed by its kinetic law, the balance becomes:

$$r'''_{CO2} = k_l a\left(CO_{2l} - CO^*_{2,m}\right) \qquad [1.25]$$

and the concentration of CO_2 dissolved in the medium is calculated by the following relation:

$$CO_{2l} = CO^*_{2,m} + \frac{r'''_{CO2}}{k_l a} \qquad [1.26]$$

As a result, the concentration of dissolved CO_2 is always greater than the solubility of CO_2. This difference increases with the volumetric rate of CO_2 production, but decreases as the transfer coefficient $k_l a$ increases.

During a nongrowth-associated metabolite production fermentation, dissolved CO_2 concentration shows a temporal variation that is similar to that of CO_2 solubility: it increases during the rapid growth phase and decreases during the metabolite production phase (Figure 1.36).

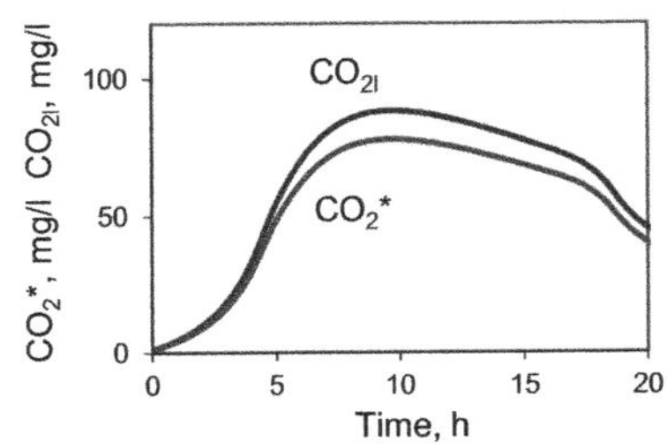

Figure 1.36. *Variation in CO_2* solubility and CO_{2l} concentration of dissolved CO_2*

1.6. References

Bailey, J.E. and Ollis, D.F. (1986). *Biochemical Engineering Fundamentals*, 2nd edition. McGraw Hill, New York.

Blanch, H.W. and Clark, D.S. (1996). *Biochemical Engineering*. Marcel Dekker, Boca Raton.

Cooney, C.L., Wang, D.I.C., Mateles, R.I. (1968). Measurement of heat evolution and correlation with oxygen consumption during microbial growth. *Biotechnology and Bioengineering*, 67(6), 269–281.

Dixon, N.M. and Kell, D.B. (1989a). The control and measurement of 'CO_2' during fermentations. *Journal of Microbiological Methods*, 10(3), 155–176.

Dixon, N.M. and Kell, D.B. (1989b). The inhibition by CO_2 of the growth and metabolism of micro-organisms. *Journal of Applied Bacteriology*, 67(2), 109–136.

Doran, P.M. (2013). *Bioprocess Engineering Principles*, 2nd edition. Academic Press, San Diego.

Felder, R.M. and Rousseau, R.W. (2005). *Elementary Principles of Chemical Processes*, 3rd edition. Wiley, New York.

Geankoplis, C.J., Hersel, A.A., Lepek, D.H. (2018). *Transport Processes and Separation Process Principles*. Prentice Hall, Upper Saddle River.

Jones, R.P. and Greenfield, P.F. (1982). Effect of carbon dioxide on yeast growth and fermentation. *Enzyme and Microbial Technology*, 4(4), 210–223.

Liu, S. (2020). *Bioprocess Engineering: Kinetics, Sustainability, and Reactor Design*, 3rd edition. Elsevier, Amsterdam.

Luong, J.H.T. and Volesky, B. (2005). Heat evolution during the microbial process: Estimation, measurement, and applications. *Microbial Activities*, 28, 1–40.

Nielsen, J. (2006). Microbial process kinetics. In *Basic Biotechnology*, Ratledge, C. and Kristiansen, B. (eds). Cambridge University Press, Cambridge.

Pouliot, K., Thibault, J., Garnier, A., Acuña Leiva, G. (2000). K L a evaluation during the course of fermentation using data reconciliation techniques. *Bioprocess Engineering*, 23, 565–573.

van't Riet, K. (1983). Mass transfer in fermentation. *Trends in Biotechnology*, 1(4), 113–119.

Shuler, M., Kargi, F., DeLisa, M. (2017). *Bioprocess Engineering: Basic Concepts*. Prentice Hall, Boston.

2

Fermenter Implementation: Principle and Optimization

Mohamed GHOUL and Jean-Marc ENGASSER
Laboratoire Réaction et Génie des Procédés (LRGP), CNRS, Université de Lorraine, Nancy, France

2.1. Optimal fermenter implementation

This chapter is dedicated to the implementation of microbial fermenters and examines the optimal selection of a fermenter's operating mode in detail. The main objective lies in identifying the most appropriate implementation among the various alternatives available, namely, batch mode, continuous mode and fed-batch mode. At the same time, we will explore how to configure the operating parameters that are specific to each mode.

Optimizing fermentation processes is one of the major challenges in bioprocess engineering. There is no universal optimization solution applicable to all types of microbial fermenters. The most appropriate implementation of a fermentation process depends on a multitude of factors, including the type of microbial transformation, the specific characteristics of the microbial cells used, notably their kinetic properties and the industrial context within which the process takes place (Atkinson and Mavituna 1991; van't Riet and Tramper 1991; Wang and Zhong 2007; Villadsen et al. 2011; Liu 2020).

Bioreactor Implementation in the Agro-Food Industries,
coordinated by Mohamed GHOUL.

In this chapter, we will outline the digital approach to fermenter optimization based on the use of simulation models (Schügerl and Bellgardt 2012; Mears et al. 2017; Koulouris et al. 2021; Du et al. 2022). These models enable us to better understand and assess the impact of various operational variables on fermenter performance. They can also be used to determine optimal values, thus guaranteeing the most efficient operation of fermenters. This digital fermenter optimization is applied to the four main operating modes: batch, continuous, fed-batch and perfusion. Its application is illustrated in current examples of aerobic and anaerobic fermentations for cell and metabolite production.

2.1.1. *Fermenter optimization criteria*

2.1.1.1. *Production cost criteria*

In this review of the optimal implementation of microbial fermentations, the optimization criterion is to minimize production costs. This economic criterion is a major factor in the industrial scale-up of fermentation processes. Fermenter optimization can also take additional criteria into account, related to product quality or plant robustness and safety.

In the context of a year-round fermentation process, the annual production cost $€_{P,year}$ takes into account the costs of raw materials, equipment, energy consumption and labor:

$$€_{p,year} = €_{mat,year} + €_{equi,year} + €_{ener,year} + €_{lab,year} \qquad [2.1]$$

The unit production cost per kilogram of product, denoted by $€_{P/kgP}$, is calculated by dividing the annual cost by the mass of product $m_{P,year}$ generated during the year:

$$€_{P/kg} = \frac{€_{p,year}}{m_{p,year}} \qquad [2.2]$$

It is this production cost per unit quantity of product that is minimized by the fermentation optimization approach.

2.1.1.2. *Technical performance criteria*

For the purposes of this chapter, we will not explicitly use production cost as an optimization criterion. Instead, we will use two technical performance indicators that largely determine production cost: fermenter productivity and substrate conversion efficiency. This approach will enable us to better understand the specific role of operational variables on fermenter operating efficiency.

2.1.1.2.1. Fermenter productivity

The first technical criterion is the fermenter's productivity, denoted as $Prod_P$ (gP/L.h or kgP/m^3.h), where the product P can be either cell biomass or a metabolite. For a fermentation process running over a year, productivity is calculated by dividing the annual mass of product $m_{p,year}$ by the useful volume of the fermenter V_f and by the annual operating time t_{year}:

$$Prod_p = \frac{m_{p,year}}{V_f t_{year}} \qquad [2.3]$$

Fermenter productivity represents the average volumetric rate of fermenter production over the year. As such, it can be seen as a measure of the efficiency with which the fermenter is used.

2.1.1.2.2. The global conversion efficiency of the substrate

The second criterion is the global conversion efficiency of the carbon substrate into product $Y_{P/S,glob}$. It is calculated by dividing the mass of product obtained at the end of fermentation by the mass of substrate used. It is also the ratio between the annual mass of product generated and the annual mass of substrate used:

$$Y_{P/S,glob} = \frac{m_{p,f}}{m_{S,}f} = \frac{m_{p,year}}{m_{S,}year} \qquad [2.4]$$

Thus defined, global conversion efficiency is a measure of the efficiency of substrate-to-product conversion within the fermenter.

It is important to note the difference between this global conversion efficiency and the $Y_{P/S}$ stoichiometric coefficient of microbial transformation previously introduced for microbial kinetics. The $Y_{P/S}$ coefficient represents the mass of product generated per gram of substrate consumed by the cell, while the global conversion efficiency $Y_{P/S,glob}$ represents the mass of product generated per gram of substrate introduced into the fermenter.

2.1.1.3. *Relationship between production costs and technical criteria*

The production cost of a fermentation depends on the two criteria of fermenter productivity and overall substrate conversion efficiency. This section establishes relationships describing the influence of the two criteria on the four contributions of material, equipment, energy and labor to the unit production cost per kilogram of product.

2.1.1.3.1. Influence of substrate conversion efficiency on material cost

For most fermentations, the carbon substrate, which is usually sugar, represents the main contribution to the cost of the substrates used. It can represent up to 90% of the cost of raw materials. As a first approximation, the annual cost of raw materials can therefore be equated with the annual cost $€_{S,year}$ of the carbonaceous substrate, which is the annual mass $m_{S,year}$ of substrate used, multiplied by the unit cost $€_{S/kgS}$ of substrate.

Under these conditions, the material cost per kilogram of product, denoted as $€_{mat/kg}$, is calculated as follows:

$$€_{mat}/_{kgP} = \frac{€_{S,year}}{m_{p,year}} = \frac{m_{S,year}€_{S}/_{kgS}}{m_{p,year}} = \frac{€_{S}/_{kgS}}{Y_{P}/_{S,glob}} \qquad [2.5]$$

According to this relationship, the cost of materials per kilogram of product is the unit cost of the carbon substrate, divided by the overall efficiency of converting the substrate into product.

The assessment of the influence of conversion efficiency on material cost can be refined by taking into account the cost of other substrates used. This is achieved by adding the cost of other substrates used per kilogram of carbon substrate to the unit cost of the substrate $€_{S/kgS}$.

2.1.1.3.2. Influence of fermenter productivity on equipment costs

The annual equipment cost is a fixed cost, corresponding to the annual depreciation and maintenance costs of the fermenter. We can define a fermenter utilization cost per cubic meter of fermenter and per hour, denoted as $€_{equip/m3.h}$. It is calculated by dividing the annual cost $€_{equip,year}$ by the volume V_f of the fermenter and by the annual operating time t_{year}:

$$€_{equip}/_{m3.h} = \frac{€_{equip,year}}{V_f t_{year}} = \qquad [2.6]$$

During the year, the production of annual product $m_{P,year}$ depends on fermenter productivity according to the following relationship:

$$m_{p,year} = Prod_p V_f t_{year} \qquad [2.7]$$

Under these conditions, the cost of equipment per kilogram of product, noted as $€_{equip/kg}$, which is the annual cost of equipment divided by the annual mass of product, is calculated as follows:

$$€_{equip}/_{kg.P} = \frac{€_{equip,year}}{m_{p,year}} = \frac{€_{equip,year}}{Prod_p V_f t_m} = \frac{€_{equip,year}/_{m3.h}}{Prod_p} \quad [2.8]$$

According to this relationship, the cost of equipment per kilogram of P is inversely proportional to fermenter productivity.

2.1.1.3.3. Influence of fermenter productivity on energy costs

A similar approach can be applied to the energy cost per kilogram of product. It takes into account the energy used to agitate the fermenter and pump fluids. We can define the energy consumption cost per cubic meter of fermenter per hour, denoted as $€_{ener/m3.h}$, and calculated from the annual energy cost by the following relationship:

$$€_{ener}/_{m3.h} = \frac{€_{ener,year}}{V_f t_{year}} \quad [2.9]$$

By introducing this quantity, the energy cost per kilogram of product, denoted as $€_{ener/kgP}$, is calculated as follows:

$$€_{ener}/_{kgP} = \frac{€_{ener,year}}{m_{p,year}} = \frac{€_{ener,year}}{Prod_p V_f t_m} = \frac{€_{ener}/_{m3.h}}{Prod_p} \quad [2.10]$$

As a result, the cost of energy consumption per kilogram of product is inversely proportional to fermenter productivity.

2.1.1.3.4. Influence of productivity on labor costs

The cost of labor (or personnel) takes into account the salaries of technicians and supervisors, as well as the company's structural costs. From the annual labor cost, we can define the labor cost per cubic meter of fermenter per hour, denoted as $€_{lab/m3.h}$, and given by the following relationship:

$$€_{lab}/_{m3.h} = \frac{€_{lab,year}}{V_f t_{year}} \quad [2.11]$$

By introducing this quantity, the cost of labor per kilogram of product, denoted as $€_{mo/kgP}$, is calculated as follows:

$$€_{lab}/_{kgP} = \frac{€_{lab,year}}{m_{p,year}} = \frac{€_{lab,year}}{Prod_p V_f t_m} = \frac{€_{lab}/_{m3.h}}{Prod_p} \qquad [2.12]$$

According to this relationship, labor costs per kilogram of product are also inversely proportional to fermenter productivity.

2.1.1.3.5. Overall influence of productivity and conversion efficiency

Based on the relationships established above, the impact of the two technical criteria on the total cost of production per kilogram of product can be described using the following relationship:

$$€_P/_{kgp} = \frac{€_S/_{kgS}}{Y_P/_{Sglob}} + \frac{€_{equip}/_{m3.h}}{Prod_p} + \frac{€_{ener}/_{m3.h}}{Prod_p} + \frac{€_{lab}/_{m3.h}}{Prod_p} \qquad [2.13]$$

According to this relationship, the minimum production cost is achieved when the fermenter operates under conditions that simultaneously promote maximum productivity and maximum substrate-to-product conversion efficiency.

However, it is frequently observed that the optimal conditions for maximizing fermenter productivity differ from those for obtaining the best substrate-to-product conversion efficiency. Optimal fermenter operation is therefore an economic compromise between the conditions required for maximum productivity and those for highest substrate conversion efficiency.

2.2. Batch fermentation

2.2.1. *Principle of batch fermentations*

In batch fermentation, the starting medium contains the nutrients required for cell growth and metabolite production. Fermentation is started by inoculating the microbial biomass. During fermentation, various elements can be introduced, such as air for aerobic microorganisms, as well as anti-foaming agents or acid/basic solutions to regulate pH. Fermentation is stopped either when one or more nutrients are depleted, or when the desired product concentration is reached.

In the context of a fermentation plant operating over periods of months or even years, batch fermentations are sequenced. Each fermentation is separated from the others by intervals of time required to empty, clean, sterilize and reload the fermenter.

Batch mode is the predominant operating method for anaerobic fermentations, such as for the production of ethanol, lactic acid, beer or waste digestion. Batch is also frequently used for aerobic fermentations to produce microbial cells and metabolites.

Batch fermentations offer the advantage of simplicity, as no nutrients are added during fermentation, reducing the risk of contamination. However, several limitations are associated with batch fermentations. These include relatively long downtimes between fermentation cycles, which reduce fermenter productivity, and the impossibility of controlling nutrient concentrations at relatively low levels.

2.2.2. *Operating variables for batch fermentations*

When using a batch fermenter, two major operational variables come into play. The first concerns the composition of the initial medium. In this layout, the initial concentration of carbon substrate, generally in the form of sugar, is specified. Concentrations of other nutrients, such as nitrogen and phosphorus, may also be taken into consideration.

The second operational variable concerns the final fermentation time. When it comes to optimizing batch fermentation, the choice is between shorter batch times, which would allow more fermentation cycles to be carried out over a given period of time, or longer batch times, resulting in fewer fermentation cycles in total. It is the final fermentation time that determines the concentration of microbial biomass or metabolites obtained.

In the case of non-growth-associated metabolite production, a third possible operational variable is the transition time between the initial phase of rapid growth and the subsequent phase of metabolite production. For example, the transition time may correspond to the introduction of a metabolite-inducing agent or an adjustment in temperature. This metabolic transition time determines the duration of the initial growth phase and, consequently, influences the concentration of active cells available for the second metabolite production phase.

2.2.3. *Batch fermentation simulation models*

The knowledge-based simulation models used here for fermenter optimization are based on the balances–thermodynamics–kinetics triangle (Figure 2.1). These involve the instantaneous material balances for the fermentation components, the kinetic laws describing the growth and metabolism of the microbial cell used, the thermodynamic law for oxygen solubility and the kinetic law for oxygen transfer between the air and the medium. These relationships were presented in the previous chapter. In this chapter, simulation models are used to assess the influence of operating variables on variations in nutrient, biomass and metabolite concentrations during fermentation.

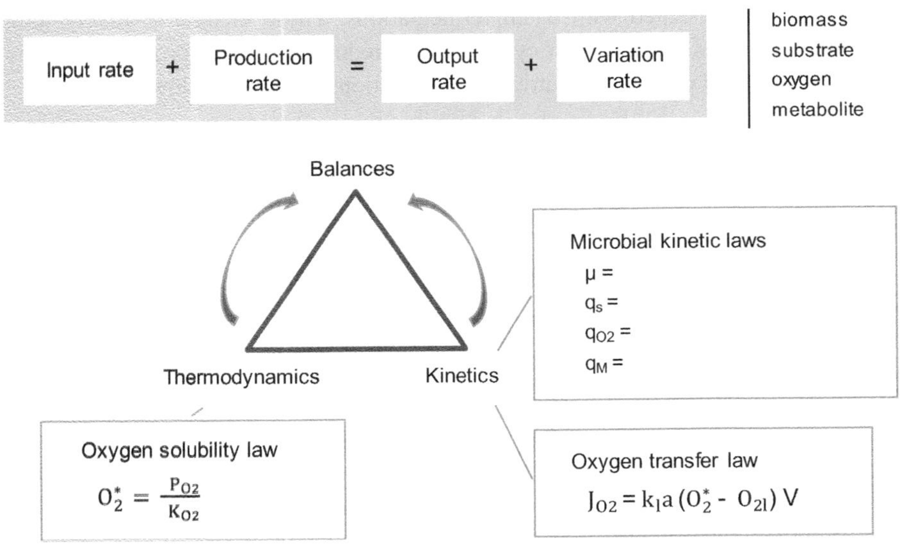

Figure 2.1. *Balance–thermodynamics–kinetics methodological triangle for modeling fermentations*

2.2.3.1. *Simulation of substrate, biomass and metabolite concentrations*

As previously explained in the context of fermentation kinetic analysis, instantaneous balances applied to biomass, substrates and metabolites can be summarized in terms of production rate and rate of change. They lead to the formulation of differential equations that translate the temporal evolution of biomass, substrate and metabolite concentrations. The balance relations and differential equations for batch mode are summarized in Figure 2.2.

To simulate fermentation, the differential equations are supplemented with kinetic laws that express the specific rates of growth, substrate consumption and metabolite production as a function of medium composition. The resulting differential equations are then solved by numerical integration, with the added substrate concentration as the initial condition.

Biomass	Substrate	Metabolite
$r_X''' = \frac{dX}{dt}$	$-r_S''' = \frac{dS}{dt}$	$r_M''' = \frac{dM}{dt}$
$\frac{dX}{dt} = \mu X$	$\frac{dS}{dt} = -q_S X$	$\frac{dM}{dt} = q_M X$

Figure 2.2. *Instantaneous biomass, substrate and metabolite balances for batch fermentation*

2.2.3.2. *Simulation of dissolved oxygen concentration*

As already mentioned in the section on oxygen, variations during the fermentation process in the mole fraction of oxygen in the fermenter output air, and in the concentration of dissolved oxygen, are governed by the two instantaneous oxygen balance equations.

The first relationship involves the instantaneous oxygen balance for the entire fermenter. This contains the oxygen input and output rates in the circulating air, as well as the oxygen consumption rate. The internal rate-of-change term is negligible, due to the very low solubility of oxygen. This balance makes it possible to assess the mole fraction of oxygen in the output air, which is then used to calculate the average partial pressure of oxygen in the fermenter, as well as the average solubility of oxygen in the medium.

The second relationship corresponds to the instantaneous oxygen balance for the liquid medium. This establishes equality between oxygen transfer and consumption rates. This equation is used to calculate the dissolved oxygen concentration in the medium, taking into account the oxygen transfer coefficient k_la in the fermenter. The relationships between the two oxygen balances are summarized in Figure 2.3.

$$32\,\frac{Q_{a,input}}{V_{m,input}}\,y_{O2,input} - r'''_{O2}\,V = 32\,\frac{Q_{a,output}}{V_{m,output}}\,y_{O2,output} \qquad [2.14]$$

$$\rightarrow y_{O2,output} = \frac{Q_{a,input}}{V_{m,input}}\,\frac{V_{m,output}}{Q_{a,output}}\,y_{O2,input} - \frac{r'''_{O2}\,V}{32\,\frac{Q_{a,output}}{V_{m,output}}} \qquad [2.15]$$

$$k_l a\,(O^*_{2,m} - O_{2l})\,V = r'''_{O2}\,V \qquad [2.16]$$

$$\rightarrow O_{2l} = O^*_{2,m} - \frac{r'''_{O2}}{k_l a} \qquad [2.17]$$

Figure 2.3. *Instantaneous oxygen balances to determine mole fraction of oxygen output and dissolved oxygen concentration*

2.2.3.3. *Calculating fermenter productivity and substrate conversion efficiency*

From the temporal variation in the composition of the medium, the model also calculates variations in the two technical performance criteria: fermenter productivity and substrate-to-product conversion efficiency.

In the case where the target product is microbial biomass, $Prod_X$ productivity is calculated as the ratio between the biomass concentration produced and the total fermentation time. This time includes both the final fermentation time and the additional time, denoted as t_{ad}, between two successive fermentation cycles.

$$Prod_X = \frac{X - X_0}{t + t_{ad}} \qquad [2.18]$$

If the target product is a metabolite, $Prod_M$ productivity is calculated by assessing the ratio between the metabolite concentration produced and the total fermentation time:

$$Prod_M = \frac{M}{t + t_{ad}} \qquad [2.19]$$

With regard to the second criterion of overall substrate conversion efficiency, if the product is microbial biomass, the conversion efficiency $Y_{X/S,glob}$ is calculated as the ratio between the concentration of biomass produced and the initially added concentration of substrate S_0:

$$Y_{X/S,glob} = \frac{X - X_0}{S_0} \qquad [2.20]$$

When the product is a metabolite, the conversion efficiency $Y_{M/S,glob}$ is calculated as the ratio between the concentration of metabolite produced and the concentration of substrate initially added:

$$Y_{M/S,glob} = \frac{M}{S_0} \qquad [2.21]$$

In this analysis of fermentation optimization, simulation models are used to identify the operational variables that achieve the highest values of fermenter productivity and overall substrate conversion efficiency.

2.2.4. *Optimization of anaerobic fermentation implementation*

The simplest optimization strategy applies to anaerobic fermentations where oxygen transfer is not involved. In typical anaerobic fermentation, sugar is converted into biomass and a metabolite. Metabolite production is closely linked to cell growth, as can be seen from examples such as ethanol and lactic acid. Under anaerobic conditions, growth rates are relatively slow. On the other hand, when excreted metabolite concentrations reach high levels, they tend to inhibit both cell growth and metabolite production.

The simulation model is used to identify the initial sugar concentration S_0 and the fermentation time t_f that lead to maximum fermenter productivity, as well as the most thorough conversion of substrate to metabolite. These two operational parameters are intrinsically linked, as a higher initial substrate concentration requires a longer fermentation time for optimal transformation. The first phase of the optimization process involves determining the fermentation time which, for a specified initial substrate concentration, results in the highest fermenter productivity.

Productivity is directly determined by fermentation kinetics, in particular by the volumetric rate of metabolite production during batch fermentation. For a relatively high initial substrate level, the rate of metabolite production increases in tandem with cell growth. However, this increase is slowed by the increasing inhibitory effect exerted by the metabolite concentration, leading to a maximum value followed by a decrease in the production rate.

As a result, fermenter productivity, which represents the average volumetric rate during fermentation, also increases to a maximum value before gradually decreasing (Figure 2.4(a)). The simulation is implemented to determine the optimal time for fermentation, when productivity is highest, and to analyze how this time evolves as a function of initial substrate concentration (Figure 2.4(b)). As expected, the time associated with maximum productivity increases in proportion to the initial substrate concentration. Note that optimized fermentation is not continued until the maximum concentration of metabolite that can be excreted by the microbial strain used is reached. Instead, fermentation is preferably stopped at an intermediate metabolite concentration that corresponds to maximum productivity.

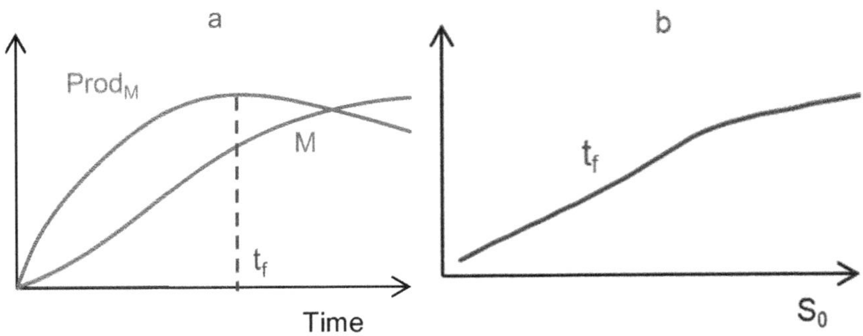

Figure 2.4. *(a) Variation during anaerobic batch fermentation in metabolite M concentration and metabolite Prod$_M$ productivity. The dotted line indicates the duration of fermentation at maximum productivity. (b) For batch fermentation, variation in fermentation time tf for maximum productivity as a function of initial substrate concentration S_0. For a color version of this figure, see www.iste.co.uk/ghoul/bioreactor.zip*

In a second step, the simulation analyzes the impact of the initial substrate concentration, both on maximum productivity at predefined fermentation times and on the overall efficiency of substrate to metabolite conversion (Figure 2.4(c)). Productivity peaks at intermediate substrate concentrations, while substrate conversion efficiency peaks at lower concentrations.

The reduction in conversion efficiency at higher substrate concentrations indicates that the substrate is no longer fully consumed after a fermentation time, leading to maximum productivity. According to the simulation, the optimal initial substrate concentration is determined to be that which generates maximum productivity together with near-maximum substrate conversion efficiency. By running the fermentation under these optimal conditions, all of the substrate is consumed, and the fermentation reaches biomass and metabolite concentrations corresponding to maximum fermenter productivity (Figure 2.4(d)).

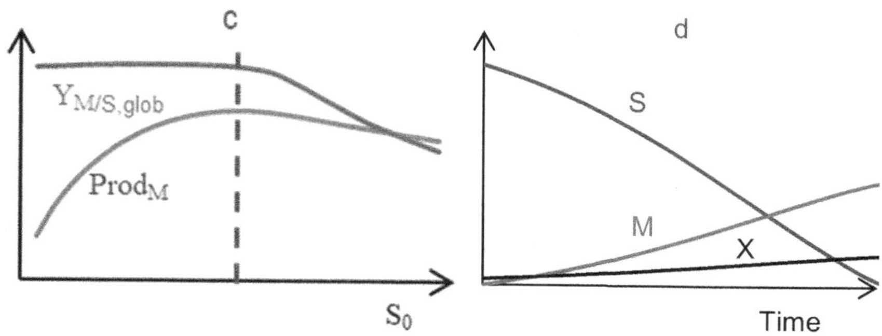

Figure 2.4. *(c) Influence of initial substrate concentration S_0 on the fermenter's productivity $Prod_M$ and global conversion efficiency of substrate into metabolite $Y_{M/S,glob}$. The dotted line indicates the substrate concentration for maximum productivity. (d) Under optimal conditions for batch fermentation, variation over time in concentrations of substrate S, biomass X and metabolite M. For a color version of this figure, see www.iste.co.uk/ghoul/bioreactor.zip*

2.2.5. *Optimization of aerobic fermentations*

In the case of aerobic fermentations designed to produce microbial biomass or metabolites, the optimization strategy is more complex due to the additional limiting step of transferring oxygen from the air to the medium. This step limits both biomass concentration and fermenter productivity.

2.2.5.1. *Concept of maximum oxygen transfer rate*

To gain a better understanding of the impact of oxygen transfer limitations on fermenter efficiency, it is worth introducing the concept of the maximum oxygen transfer rate. The volumetric rate at which oxygen is transferred between the air and the medium, denoted as J_{O2}''' and expressed in $gO_2/L.h$, is defined by the following kinetic law:

$$J_{O2}''' = k_l a \left(O_{2,m}^* - O_{2l}\right) \qquad [2.22]$$

For a given aeration capacity k_la, the volumetric rate of oxygen transfer reaches a maximum value when the difference between the mean solubility $O_2{}^*{}_m$ and the dissolved oxygen concentration O_{2l} is highest; in other words, when O_{2l} reaches zero. However, in reality, it is preferable to maintain the dissolved oxygen concentration above a critical minimum value $O_{2l,min}$ in order to prevent a reduction in growth rate or an accumulation of undesirable metabolites. In this context, the maximum oxygen transfer rate is reached when the dissolved oxygen concentration is equal to $O_{2l,min}$. Consequently, the maximum oxygen transfer rate, denoted as $J'''_{O2,max}$, is expressed by the following relationship:

$$J'''_{O2,max} = k_l a\left(O^*_{2,m} - O_{2l,min}\right) \qquad [2.23]$$

The maximum rate of oxygen transfer varies during fermentation due to variations in the average solubility of oxygen resulting from variations in the mole fraction of oxygen leaving the fermenter.

To better visualize the role of oxygen transfer limitations, it is worth plotting temporal variations in the volumetric rate of oxygen consumption and the maximum rate of oxygen transfer on the same graph. Fermentation is not affected by oxygen transfer as long as the oxygen consumption rate remains below the maximum oxygen transfer rate. Under these conditions, the dissolved oxygen concentration remains above its minimum critical value (Figure 2.5).

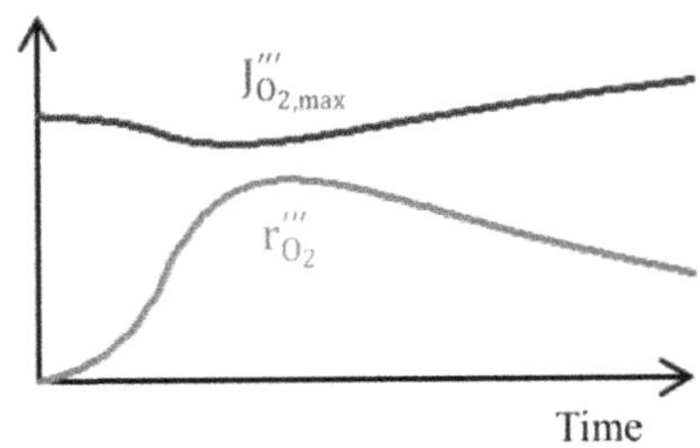

Figure 2.5. *Variation of the volumetric rate of oxygen consumption r'''_{O2} and the maximum volumetric rate of oxygen transfer $J'''_{O2,max}$ during aerobic fermentation. For a color version of this figure, see www.iste.co.uk/ghoul/bioreactor.zip*

2.2.5.2. *Maximum cell concentration through oxygen transfer*

Using the concept of maximum oxygen transfer rate, it is possible to determine the maximum biomass concentration achievable before oxygen limitations become apparent. Since the volumetric rate of oxygen consumption increases as the biomass concentration increases, the maximum cell concentration X_{max} is reached when the oxygen consumption rate is at its maximum, equal to the maximum oxygen transfer rate [2.20].

$$J'''_{O2,max} = r'''_{O2,max} = Y_{O2/X}\, \mu\, X_{max} \qquad [2.24]$$

or when: [2.21]

$$k_l a\left(O^*_{2,m} - O_{2l,min}\right) = Y_{O2/X}\,\mu\, X_{max} \qquad [2.25]$$

Consequently, the maximum biomass concentration allowed by oxygen transfer is determined by the following relationship:

$$X_{max} = \frac{k_l a\,(O^*_{2,m} - O_{2l,min})}{Y_{O2/X}\,\mu} \qquad [2.26]$$

Figure 2.6 shows a simulation of expected X_{max} values as a function of specific growth rate and oxygen transfer coefficient.

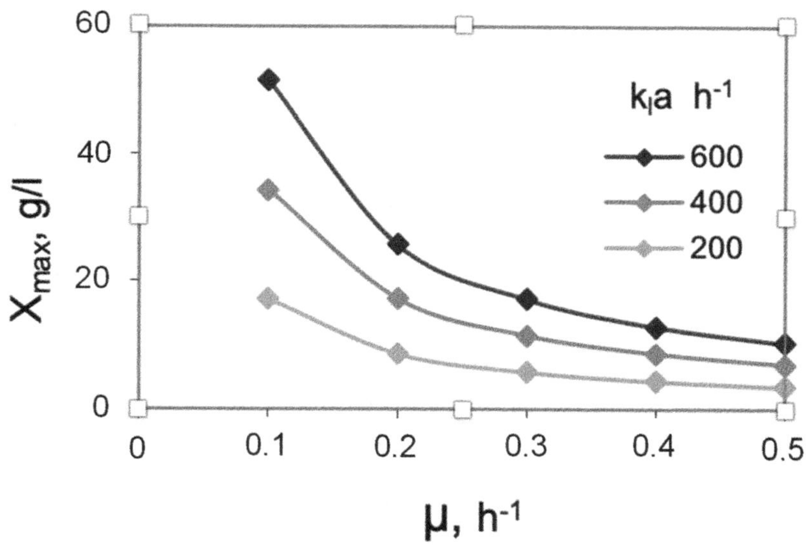

Figure 2.6. *Influence of specific growth rate μ and oxygen transfer coefficient $k_l a$ on the maximum cell concentration allowed by oxygen transfer*

The figure shows that by increasing the specific growth rate from 0.1 to 0.5 h^{-1}, X_{max} decreases from 17 to 3 g/L at a $k_l a$ of 200 h^{-1}, for example, using air at a pressure of 1 atm. The maximum biomass concentration can be increased by increasing the $k_l a$ of the fermenter or the partial pressure of oxygen in the input air.

2.2.5.3. *Optimization of batch fermentation for biomass production*

The optimization of aerobic batch fermentation is first illustrated for biomass production from a carbon substrate, with the concurrent possibility of generating metabolites. The first step in the optimization process is to determine the fermentation time which, at a specified initial substrate concentration, leads to maximum fermenter productivity. For this aerobic fermentation, the optimal fermentation time corresponds to the point at which the volumetric rate of oxygen consumption reaches the maximum oxygen transfer rate and dissolved oxygen reaches its critical minimum concentration (Figure 2.7(a)). This is also when the biomass concentration reaches the maximum value allowed by the oxygen transfer capacity of the fermenter. This optimal fermentation time increases with the substrate concentration that was initially added to the medium (Figure 2.7(b)).

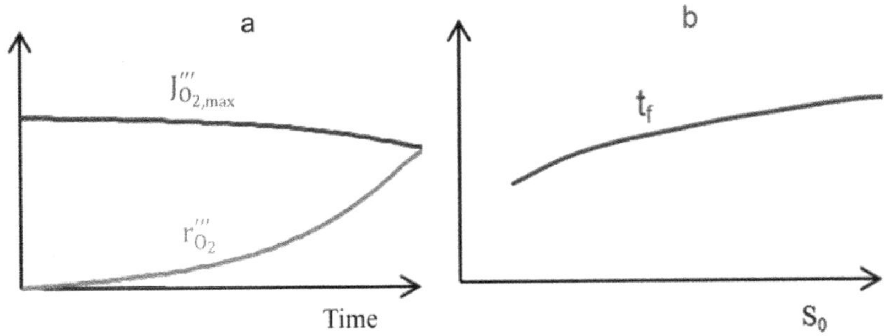

Figure 2.7. *(a) Variation of oxygen consumption volumetric rate r'''_{O2} and maximum oxygen transfer volumetric rate $J'''_{O2,max}$ during aerobic batch cell production. Fermentation is stopped when the two rates meet. (b) For batch fermentation, influence of initial substrate concentration S_0 on fermentation time t_f to reach maximum cell concentration for oxygen transfer. For a color version of this figure, see www.iste.co.uk/ghoul/bioreactor.zip*

As with anaerobic fermentation, in a second step, the simulation examines the influence of initial substrate concentration on both maximum productivity at the times previously determined, and on overall substrate-to-biomass conversion efficiency (Figure 2.7(c)). The results show a progressive increase in productivity with substrate concentration, until a peak is reached corresponding to the maximum biomass concentration. At the same time, substrate conversion efficiency remains constant at its highest level. As a result, the optimal initial substrate concentration is similar to the one corresponding to maximum fermenter productivity. When batch fermentation is carried out under these optimal conditions, biomass production continues until the dissolved oxygen level drops to its minimum value. The substrate is then completely consumed, and fermentation reaches the biomass concentration corresponding to maximum fermenter productivity (Figure 2.7(d)).

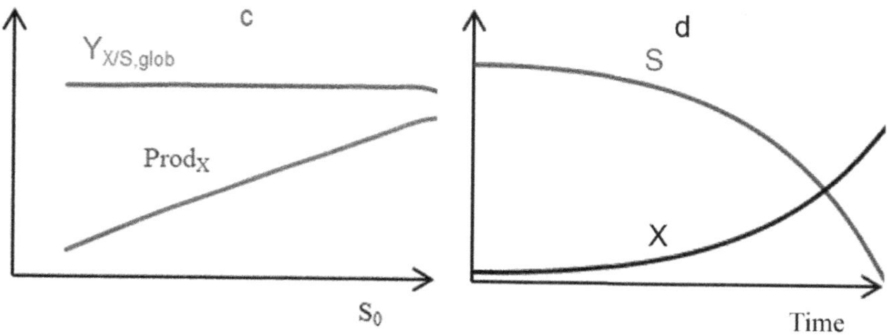

Figure 2.7. *(c) For batch fermentation, influence of initial substrate concentration S_0 on $Prod_X$ fermenter productivity allowed by aeration, and global conversion efficiency of substrate into metabolite $Y_{M/S,glob}$. (d) Under optimal batch fermentation conditions, variation in substrate and biomass concentrations over time. For a color version of this figure, see www.iste.co.uk/ghoul/bioreactor.zip*

2.2.5.4. *Optimization of batch fermentation for metabolite production*

A second example of optimizing aerobic fermentation concerns the production of a metabolite that is not associated with growth. Production of the metabolite can be triggered by the addition of a specific inducer, or by a change in either the temperature or pH of the medium.

During batch fermentation, the volumetric rate of oxygen consumption increases during the initial phase of rapid growth, reaches a maximum level at the end of growth, and then decreases during metabolite production. Simultaneously, dissolved oxygen concentration initially decreases until it reaches a minimum value, then gradually increases.

The first step in the fermentation optimization strategy is to determine the transition time between the growth and production phases (Figure 2.8(a)). The aim is to achieve the highest possible cell concentration, which is made possible by the oxygen transfer rate. The optimal transition time is evaluated as the time at which the peak of the oxygen consumption rate coincides with the maximum oxygen transfer rate. It is at this precise moment that the maximum cell concentration can be obtained (Figure 2.8(b)).

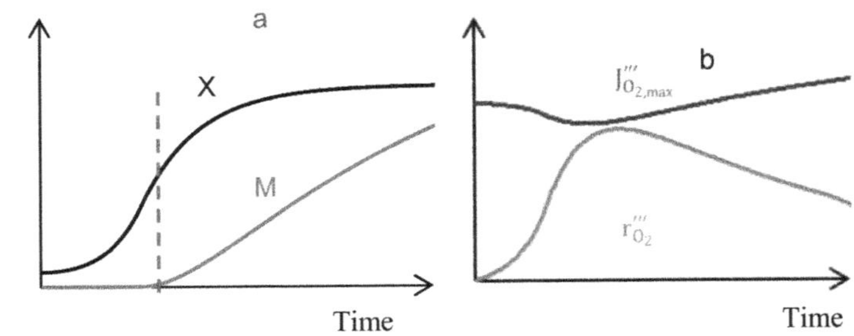

Figure 2.8. *(a) Variation of non-growth-associated metabolite in the concentrations of biomass X and metabolite M during aerobic production. The dotted line indicates the induction time of metabolite production. (b) Variation of oxygen consumption volumetric rate r'''_{O2} and maximum oxygen transfer volumetric rate $J'''_{O2,max}$ under maximum cell production conditions, during batch fermentation. For a color version of this figure, see www.iste.co.uk/ghoul/bioreactor.zip*

In a second step, fermentation duration is assessed, with the aim of achieving maximum fermenter metabolite productivity (Figure 2.8(c)). Fermenter productivity, expressing the average volumetric rate of metabolite production during fermentation, increases after the transition to a maximum value, then gradually decreases. Simulation is used to identify the fermentation time at which productivity is highest, and to analyze the variation of this time as a function of initial substrate concentration (Figure 2.8(d)). With this in mind, fermentation is not continued until the maximum concentration of metabolite that can be excreted by the microbial strain used is reached. Instead, it is stopped at an intermediate metabolite concentration that corresponds to the highest productivity.

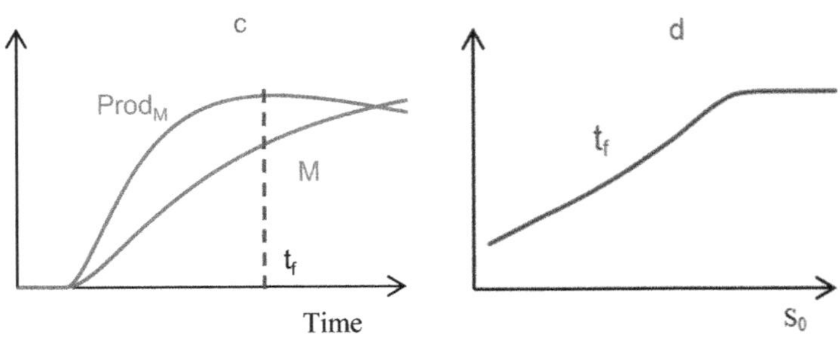

Figure 2.8. *(c) Variation in metabolite M concentration and $Prod_M$ metabolite productivity during batch fermentation. The dotted line indicates the duration of fermentation at maximum metabolite productivity. (d) Influence of initial substrate concentration S_0 on fermentation time t_f for maximum productivity. For a color version of this figure, see www.iste.co.uk/ghoul/bioreactor.zip*

In a third step, the simulation assesses the influence of the initial substrate concentration, both on productivity at the previously determined times and on the overall substrate-to-metabolite conversion efficiency (Figure 2.8(e)). This analysis shows that fermenter productivity increases as the initial substrate concentration rises, until it reaches a maximum value. However, this dependence on substrate concentration is more complex for substrate-to-metabolite conversion efficiency. At low initial substrate concentrations, the conversion efficiency decreases. This is due to the relatively high consumption of substrate to generate biomass, leaving less substrate available for metabolite production. Conversion efficiency peaks at intermediate substrate concentrations.

Conversely, at higher initial substrate concentrations, conversion efficiency decreases, suggesting that the substrate is no longer fully utilized by the time maximum productivity is recorded.

According to the simulation, the optimal initial substrate concentration is that which ultimately generates the maximum substrate conversion efficiency, together with maximum fermenter productivity. When fermentation takes place under these optimal conditions, the first phase leads to the maximum cell concentration allowed by the aeration capacity of the fermenter (Figure 2.8(f)). Then, after induction of metabolite synthesis, fermentation continues until the final metabolite concentration is reached, corresponding to maximum fermenter productivity.

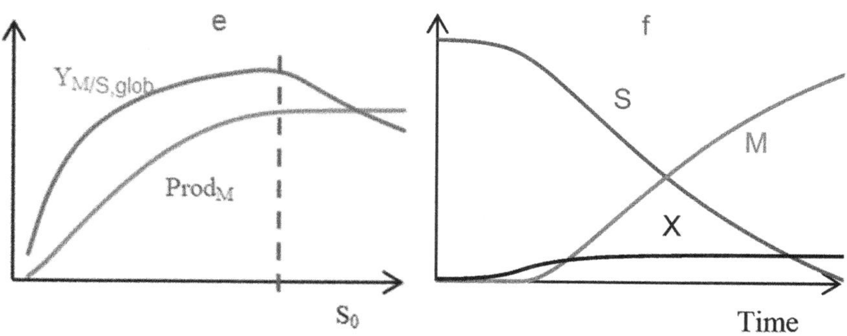

Figure 2.8. *(e) Influence of initial substrate concentration S_0 on aeration-induced fermenter productivity $Prod_M$ and global conversion efficiency of substrate into metabolite $Y_{M/S,glob}$. (f) Under optimal batch fermentation conditions, variation over time of substrate S, biomass X and metabolite M concentrations. For a color version of this figure, see www.iste.co.uk/ghoul/bioreactor.zip*

2.3. Continuous fermentations

2.3.1. *Principle of continuous fermentations*

Continuous-mode fermentation generally begins with an initial batch fermentation phase, during which the microorganisms adapt to their new environment and initiate growth. Subsequently, the fermentation medium is continuously fed into the fermenter, at the same time as the internal medium is extracted. This output medium contains unconsumed substrates, microbial cells and the metabolites generated. Feeding rates and withdrawing rates are identical, so as to maintain a constant volume of medium inside the fermenter. By implementing this method of medium feeding, the aim is to achieve stabilized fermenter operation, characterized by constant concentrations of substrates, cells and products inside the fermenter and in the output flow (Figure 2.9).

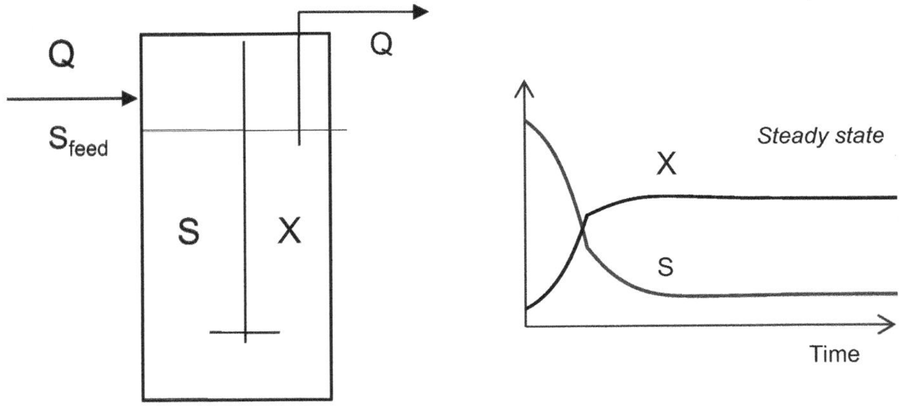

Figure 2.9. *Operating variables for continuous fermentation and variation over time in substrate S and biomass X concentrations*

Continuous fermentations offer several advantages over batch fermentations. First, they offer the possibility of running prolonged fermentation cycles that can extend over weeks (or even months). This results in a significant reduction in downtime and in the initial propagation phases of batch operation. This leads to higher overall fermenter productivity, and potentially to savings in labor costs. Furthermore, the continuous fermentation mode guarantees superior consistency in terms of final product composition and quality.

Another key advantage of continuous fermentation is its ability to operate with low limiting substrate concentrations. These substrates, which typically include sources of carbon, but also nitrogen or other elements, can be maintained at reduced levels throughout the continuous feeding of the medium. This approach can induce beneficial changes in cell metabolism. For example, it may result in reduced excretion of undesirable metabolites or enhanced production of desired metabolites (Gordeeva et al. 2021; Dong et al. 2023).

Continuous fermentations, while offering these advantages, also present significant challenges and risks. One of these main risks is the potential for microbial contamination. When a fermenter is fed continuously for long periods, the risk of contamination considerably increases. In addition, prolonged exposure of microbial cells to substrate-limiting conditions can produce physiological and metabolic variations, leading to instabilities for continuous fermentation.

Another limitation of continuous fermentation is its reduced adaptability to produce non-growth associated metabolites. Unlike batch fermentation, where distinct phases of growth and metabolite production are carried out sequentially, continuous fermenters require the simultaneous implementation of these phases. This challenge becomes particularly complex for cells excreting metabolites at low growth rates or under conditions of arrested growth.

Due to their inherent constraints and increased susceptibility to contamination, continuous fermentations are less commonly used than batch fermentations. Continuous implementations are mainly used in situations where non-sterile conditions prevail, such as wastewater treatment, bioremediation of pollutants or anaerobic digestion of waste. Continuous fermentation can also become an attractive prospect when used in selective environments, such as low pH media, which limit the proliferation of contaminants. This approach is relevant in certain applications that are designed to produce microbial cells or metabolites associated with growth.

2.3.2. *Operating variables for continuous fermentations*

Continuous fermentation involves a number of operational variables related to the consecutive phases of initial batch fermentation and continuous medium feeding. For the first phase of batch fermentation, these variables are the initial composition of the medium, in particular the concentration of the carbon source, as well as the duration of the batch operation, which generally extends over a few hours. This period determines the biomass and substrate concentrations at the start of the continuous operation. As will be detailed later, the initial batch phase influences the dynamics of the continuous operation, but has no effect on the stabilized biomass and substrate concentrations that can be achieved.

The two main operating variables concern the period of continuous medium feeding. First, the feed flow rate, which is generally kept constant, and second, the composition of the feed medium, with particular attention paid to the carbon substrate concentration. The feed flow rate determines the residence time of the medium components inside the fermenter. On the other hand, the concentration of carbon substrate in the feed has a major influence on the levels of biomass and metabolites that can be achieved.

There is no theoretical limit to the total duration of continuous fermentation. It can extend over several weeks, months, and in some cases even longer, particularly in the case of water and waste treatment. The length of time depends on the stability of the microbial cells throughout the continuous operation, and their susceptibility to contamination.

2.3.3. *Simulation model for continuous fermentations*

Compared to the model that was previously established for batch fermentation, simulation models for continuous fermentations involve the same laws for microbial kinetics, oxygen solubility and oxygen transfer rate, but have modified instantaneous balances for the components of microbial transformation, considering continuous feeding and extraction from the medium.

2.3.3.1. *Instantaneous biomass, metabolite and substrate balances*

During continuous operation, the instantaneous biomass balance involves the additional biomass output rate term. At a constant flow rate Q in L/min, the balance is expressed as follows:

$$V r_X''' = Q X + \frac{d(V X)}{dt} \qquad [2.27]$$

or

$$r_X''' = \frac{Q}{V} X + \frac{dX}{dt} \qquad [2.28]$$

Temporal variation in biomass concentration is therefore governed by the following differential equation:

$$\frac{dX}{dt} = \mu X + \frac{Q}{V} X \qquad [2.29]$$

Similar relationships apply to a metabolite produced during fermentation.

As far as the substrate is concerned, the instantaneous balance includes the two additional terms of input and output rates in the medium circulating through the fermenter. With a S_{feed} concentration in the feed, this is written as:

$$Q\,S_{feed} - V r_S''' = QS + \frac{d(V\,S)}{dt} \qquad [2.30]$$

or

$$r_S''' = \frac{Q}{V}(\,S_{feed} - S) - \frac{dS}{dt} \qquad [2.31]$$

Thus, the temporal variation in substrate concentration is determined by the following differential equation:

$$\frac{dS}{dt} = -\,q_S X + \frac{Q}{V}(\,S_{feed} - S) \qquad [2.32]$$

2.3.3.2. *Instantaneous oxygen balances*

Compared to the batch process, the instantaneous oxygen balance over the entire fermenter includes the two additional terms of input and output rates in the medium circulating through the fermenter. Assuming that the feeding medium is saturated with oxygen, the balance relationship is as follows:

$$32\frac{Q_{a,input}}{V_{m,input}}\, y_{O2,input} + Q\,O^*_{2,feed} - r_{O2}''' V = A \qquad [2.33]$$

$$A = 32\frac{Q_{a,output}}{V_{m,output}}\, y_{O2,output} + QO_{21} + \frac{dm_{O2}}{dt} \qquad [2.34]$$

Due to the very low solubility of oxygen, these two input and output rate terms, as well as the rate-of-change term, are negligible compared to the oxygen input and output rates in the air. Consequently, the oxygen balance on the continuous fermenter can be assumed to be identical to the balance on the batch fermenter:

$$32\frac{Q_{a,input}}{V_{m,input}}\, y_{O2,input} - r_{O2}''' V = 32\frac{Q_{a,output}}{V_{m,output}}\, y_{O2,output} \qquad [2.35]$$

The instantaneous oxygen balance of the liquid medium, which stipulates that the oxygen transfer rate is equal to the oxygen consumption rate, is also the same as for batch fermentation:

$$k_l a\left(O_{2,m}^{*} - O_{21}\right) = r_{O2}^{\prime\prime\prime} \quad [2.36]$$

2.3.4. *Continuous fermentation dynamics*

2.3.4.1. *Dilution rate and residence time in the continuous fermenter*

According to the instantaneous biomass and substrate balances, the temporal variations in biomass and substrate concentrations, during continuous fermentation, depend on the ratio between the medium flow rate and the internal volume of the medium. This ratio is known as the dilution rate. The dilution rate D, expressed in units h^{-1}, is given by the following relationship:

$$D = \frac{Q}{V} \quad [2.37]$$

This represents the volume of medium in liters fed per liter of internal volume, per hour.

The dynamics of continuous fermentation can also be studied by analyzing the residence time of the circulating medium within the fermenter. The residence time τ, which is the opposite of the dilution rate D, is defined as the ratio between the internal volume of the medium and the flow rate of the medium. In a continuously mixed fermenter, there is a wide range of residence times for fluid elements inside the fermenter. The residence time τ is the average residence time of all incoming fluid elements. When analyzing fermentation kinetics in terms of microbial reaction progress time, the real time scale of the batch is replaced by the residence time scale inside the continuous fermenter.

2.3.4.2. *Influence of dilution rate on fermentation dynamics*

During the initial phase of batch fermentation, substrate concentrations decrease, while cell biomass concentrations increase. When medium feeding starts, substrate and biomass concentrations can assume different temporal evolutions, depending on the dilution rate and substrate concentration in the feed medium.

As feeding continues, these concentrations reach stable states characterized by constant values (Figure 2.10). Stabilized concentrations vary according to the dilution rate of the medium. At low dilution rates, corresponding to high residence times, we observe a reduced stabilized substrate concentration, accompanied by a high biomass concentration. At intermediate dilution rates, fermentation dynamics lead to intermediate stationary concentrations of substrate and biomass. However, at excessively high feed flow rates, total elimination of cells in the fermenter is observed at the steady state. This is because, at excessively high feed flow rates, the rate at which cells leave the fermenter exceeds the rate at which cells reproduce inside, resulting in the phenomenon of cell washout.

It should be noted that the initial batch fermentation phase influences substrate and biomass concentrations during the transient period, but does not affect final steady-state biomass and substrate concentrations.

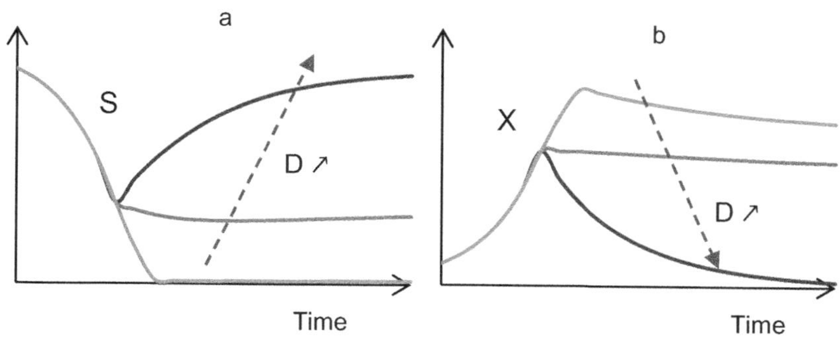

Figure 2.10. *(a) Influence of dilution rate D on the variation of substrate concentration S during continuous fermentation. (b) Influence of dilution rate D on the variation in biomass concentration X during continuous fermentation*

2.3.5. *Steady state of continuous fermentations*

2.3.5.1. *Steady-state balance relations*

The substrate and biomass concentrations reached at the steady state can be predicted from the instantaneous steady-state balances when the rate-of-change term is set to zero.

The stationary biomass concentration X_{ss} is determined by the biomass balance relation, according to which the cell production rate is equal to the cell output rate:

$$r_X''' = \frac{Q}{V} X_{ss} = DX_{ss} \quad [2.38]$$

When the volume growth rate r_X''' is expressed as a function of the specific growth rate μ, the balance becomes:

$$\mu X_{ss} = DX_{ss} \qquad [2.39]$$

or

$$\mu = D \qquad [2.40]$$

Consequently, at the steady state, the specific growth rate of the cells is equal to the dilution rate of the fermenter. This key relationship of continuous fermentation indicates that the steady state is reached when the cells have adjusted their growth rate to the dilution rate.

For a produced metabolite, the steady-state concentration M_{ss} is governed by the following balance relation:

$$r_M''' = \frac{Q}{V} M_{ss} \qquad [2.41]$$

or

$$q_M X_{ss} = DM_{ss} \qquad [2.42]$$

Furthermore, the stationary substrate concentration S_{ss} is determined by the stationary substrate balance:

$$\frac{Q}{V} S_{feed} - r_S''' S_{ss} = \frac{Q}{V} Sss \qquad [2.43]$$

or

$$q_S X_{ss} = D(S_{feed} - S_{ss}) \qquad [2.44]$$

2.3.5.2. *Influence of dilution rate on stationary substrate and biomass concentrations*

Stabilized biomass and substrate concentrations, expressed by steady-state balance relationships, depend on the kinetic laws of microbial transformation. In this section, concentrations are simply assessed by a Monod growth law, characterized by the two constants μ_{max} and K_s, and for a substrate consumption rate comprising a growth-associated term, characterized by the stoichiometric coefficient $Y_{S/X}$, and a maintenance rate m_s.

2.3.5.2.1. Stationary substrate concentration

The stationary substrate concentration is determined from the biomass balance. Expressing the specific growth rate by Monod's law, the instantaneous biomass balance becomes:

$$\frac{\mu_{max}\, S_{SS}}{K_S + S_{SS}} = D \qquad [2.46]$$

or

$$\mu_{max}\, S_{SS} = DK_S + D\, S_{SS} \qquad [2.47]$$

Consequently, the stabilized substrate concentration is calculated as a function of the dilution rate using the following relationship:

$$S_{SS} = \frac{D\, K_S}{\mu_{max} - D} \qquad [2.48]$$

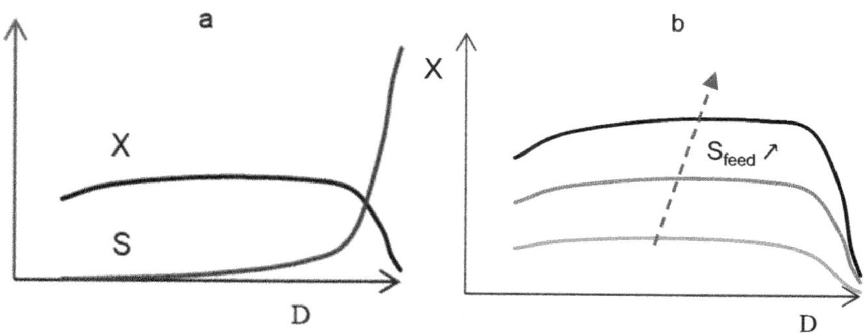

Figure 2.11. *(a) At steady state, influence of dilution rate D on substrate S and biomass X concentrations. (b) Influence of substrate concentration in the feed S_{feed} on the variation curve of stationary biomass concentration X as a function of dilution rate D*

According to this relationship, for very low dilution rates, representing long residence times of the medium in the fermenter, the substrate is fully metabolized before leaving the fermenter (Figure 2.11(a)). As the dilution rate increases, which corresponds to shorter residence times, the stationary substrate concentration also increases, eventually reaching the feed concentration level under conditions where cell washout occurs. It is important to note that the stationary substrate concentration does not depend on the substrate concentration in the feed. Substrate concentration adjusts automatically, thanks to fermentation dynamics, until the specific cell growth rate becomes equal to the dilution rate.

2.3.5.2.2. Stationary biomass concentration

The stationary biomass concentration is determined from the substrate balance. By expressing the specific substrate consumption rate in terms of its two-term kinetic law, the instantaneous substrate balance becomes:

$$\left(Y_{S/X}\mu + m_S\right) X_{ss} = D(S_{feed} - S_{ss}) \qquad [2.49]$$

Consequently, the stabilized biomass concentration is calculated as a function of the dilution rate using the following relationship:

$$X_{ss} = \frac{D}{Y_{S/X}\, D + ms}(S_{feed} - S_{ss}) \qquad [2.50]$$

According to this relationship, the stationary biomass concentration increases as a function of substrate concentration in the feed medium (Figure 2.11(b)). However, this dependence is more complex when considered in relation to dilution rate. Indeed, it reaches its maximum level at intermediate dilution rates, but tends to decrease as dilution rates become very low, mainly due to the significant consumption of substrate required for cell maintenance. As dilution rates increase, corresponding to reduced residence times, the stationary biomass concentration decreases. This reaches zero when the dilution rate becomes too high, indicating that the rate at which cells leave the fermenter outweighs their rate of reproduction within it.

When cell washout occurs, the substrate concentration inside the fermenter is the feed substrate concentration. Under these conditions, the washout dilution rate $D_{washout}$ is equal to the specific cell growth rate at the feed substrate concentration; in other words:

$$D_{washout} = \frac{\mu_{max} S_{feed}}{K_S + S_{feed}} \qquad [2.51]$$

In most cases, S_{feed} is much higher than the constant K_S, in such a way that the cell washout dilution rate is close to the maximum cell growth rate μ_{max}. As a result, continuous cell production cannot be stabilized at a dilution rate that is higher than the maximum cell growth rate.

2.3.5.3. *Substrate productivity and conversion efficiency for continuous fermentation*

2.3.5.3.1. Fermenter productivity

For continuous-mode fermentation, the cell productivity $Prod_X$ is, at a given time, calculated as the total mass of cells produced since the start of fermentation divided by the useful volume of the fermenter and the total duration of fermentation, that is to say:

$$Prod_X = \frac{V(X - X_0) + m_{X,rem}}{V(t + t_{ad})} \quad [2.52]$$

The mass of cells produced is made up, on the one hand, of the increase in cell mass inside the fermenter, from the initial concentration X_0 to concentration X and, on the other hand, it includes the cell mass removed from the fermenter in the output flow rate $m_{X,rem}$. This latter quantity is calculated by integrating the cell output rate according to the following relationship:

$$m_{X,rem} = \int_0^t Q\,X\,dt \quad [2.53]$$

As the steady state is approached, fermenter productivity leans toward the stabilized biomass production volumetric rate, which, according to the biomass balance, is equal to the output volumetric rate of the fermenter cells:

$$Prod_X = \frac{Q\,X_{ss}}{V} = D\,X_{ss} \quad [2.54]$$

In the case of Monod growth kinetics, by introducing the relationship of steady-state biomass concentration X_{ss} as a function of dilution rate D, we obtain the equation for productivity as a function of D:

$$Prod_X = \frac{D^2\,Y_{X/S}}{D + m_S Y_{X/S}}\left(S_{feed} - \frac{D\,K_S}{\mu_{max} - D}\right) \quad [2.55]$$

The curve representing the variation of productivity $Prod_X$ as a function of D adopts a bell shape (Figure 2.12(a)). Its peak is reached for a dilution rate value that is relatively close to the washout dilution rate. At the lowest and highest dilution rates, productivity decreases. On the other hand, fermenter productivity increases with substrate concentration in the feed, correlating with the increase in stationary biomass concentration.

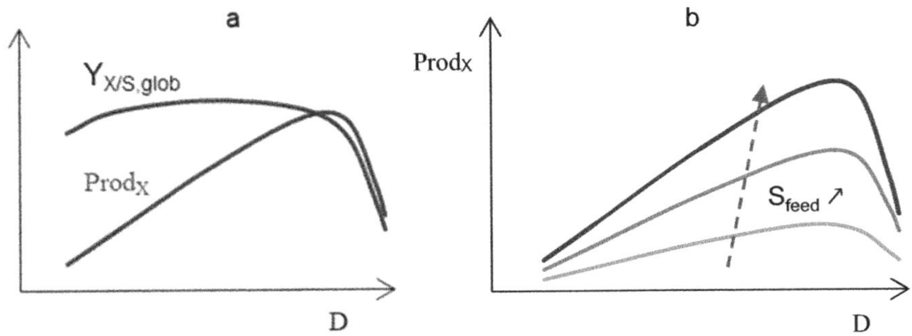

Figure 2.12. *(a) At steady state, influence of dilution rate D on fermenter biomass productivity $Prod_X$ and global conversion efficiency of the substrate into biomass $Y_{X/S}$. (b) Influence of substrate concentration in the feed S_{feed} on the variation curve of productivity $Prod_X$ as a function of dilution rate D*

2.3.5.3.2. Global substrate-to-biomass conversion efficiency

The global conversion efficiency of substrate into biomass $Y_{X/S,glob}$ is calculated as the mass of cells produced at the end of fermentation divided by the total mass of substrate used, which is the substrate initially introduced and the substrate $m_{S,added}$ added during fermentation:

$$Y_{X/S,glob} = \frac{V(X - X_0) + m_{X,rem}}{V\,S_0 + m_{S,added}} \qquad [2.56]$$

The amount of substrate added is calculated as follows:

$$m_{S,added} = \int_0^t Q\,S_{feed}\,dt \qquad [2.57]$$

When steady state is reached, conversion efficiency converges to the ratio of biomass production rate to substrate input rate; in other words:

$$Y_{X/S\,glob} \frac{Y_X}{S}, glob = \frac{D}{Y_{S/X}\,D + m_S} \qquad [2.58]$$

Introducing the expression of the stationary biomass concentration X_{ss} as a function of the feed concentration S_{feed}, and considering that the residual substrate concentration is negligible when compared to the feed concentration, the overall substrate conversion efficiency can be expressed as follows:

$$\frac{Y_X}{S},_{glob} = \frac{D}{Y_{S/X}\, D + m_S} \qquad [2.59]$$

According to this equation, at steady state, substrate conversion efficiency does not depend on substrate feed concentration (Figures 2.12(a) and (b)). It is at its maximum over the whole range of D values, when the residual substrate concentration is very low. However, at high dilution ratios, conversion efficiency decreases simultaneously with increasing substrate concentration.

2.3.6. *Optimization of continuous anaerobic fermentation for metabolite production*

The implementation of continuous anaerobic fermentation represents a technically attractive option for the production of a growth-associated metabolite. The fermentation considered is similar to that which was previously analyzed for batch fermentation optimization. It converts sugar into biomass and a growth-associated metabolite. From a kinetic point of view, growth and metabolite production rates are inhibited at high metabolite concentrations.

In the steady state, biomass and metabolite concentrations increase simultaneously with a decrease in the dilution rate, while the stationary substrate concentration decreases (Figure 2.13(a)). Comparing this substrate concentration variation curve with the one that was previously obtained for Monod growth kinetics, we observe that substrate concentration decreases less rapidly as the dilution rate is reduced. This difference is due to inhibition of the growth rate by the metabolite. Indeed, in the presence of this inhibition, a higher substrate concentration is required to achieve a specific growth rate that is equal to the dilution rate.

The continuous fermentation simulation model is used to identify the two main operating variables, namely, the dilution rate and the substrate feed concentration, which result in the highest productivity in the fermenter, as well as the highest possible conversion of substrate to metabolite. The first phase of the optimization procedure aims to determine the dilution rate which, at a specified substrate feed concentration, generates the highest fermenter productivity. According to the simulation results, fermenter productivity, which, at a steady state, represents the volumetric rate of metabolite production, peaks at an intermediate dilution rate (Figure 2.13(b)). This productivity decreases at higher dilution rates, due to the drop in cell concentration. Similarly, it decreases at lower dilution rates, due to the inhibition caused by high metabolite concentrations.

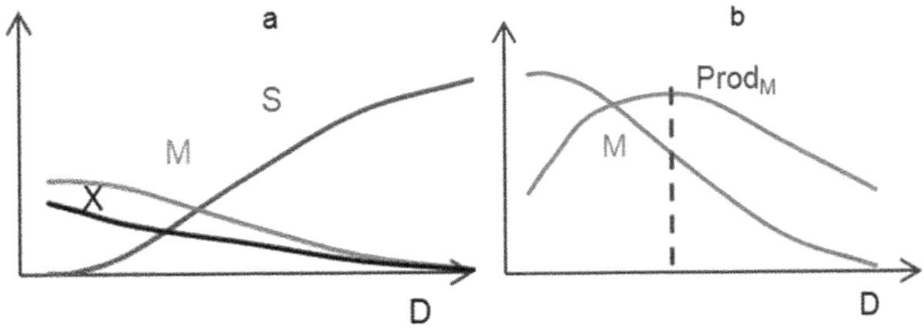

Figure 2.13. *(a) For anaerobic fermentation of growth-associated metabolite, influence of dilution rate D on stationary concentrations of substrate S, biomass X and metabolite M. (b) Influence of dilution rate D on the stationary concentration of metabolite M and fermenter productivity of metabolite $Prod_M$. The dotted line indicates the dilution speed for maximum productivity. For a color version of this figure, see www.iste.co.uk/ghoul/bioreactor.zip*

The simulation allows us to identify the dilution rate for which productivity is highest, and to determine how this dilution rate varies as a function of substrate concentration in the feed. We observe that the dilution rate that maximizes productivity decreases as the substrate concentration in the feed increases (Figure 2.13(c)). It should be noted that optimized fermentation does not aim to achieve the maximum concentration of metabolite that the microbial cell can excrete. On the contrary, continuous fermentation is stabilized at an intermediate metabolite concentration that corresponds to maximum fermenter productivity.

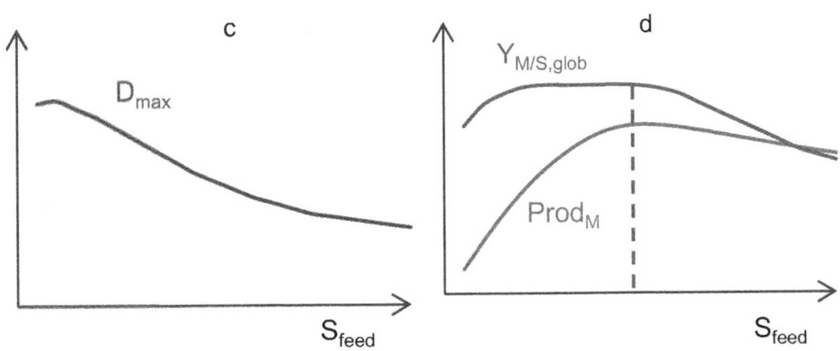

Figure 2.13. *(c) Influence of substrate concentration in the feed S_{feed} on the dilution rate D_{max} at which metabolite productivity is at its maximum. (d) Influence of the substrate concentration in the feed S_{feed} on the productivity of the $Prod_M$ fermenter and the global conversion efficiency of substrate into metabolite $Y_{M/S,glob}$. The dotted line indicates the S_{feed} concentration at which productivity is at its maximum. For a color version of this figure, see www.iste.co.uk/ghoul/bioreactor.zip*

In a second step, the simulation assesses the influence of feed concentration, both on the highest productivity at the previously determined dilution rate and on the global conversion efficiency of substrate into metabolite (Figure 2.13(d)). It indicates that fermenter productivity reaches a maximum value at an intermediate substrate concentration. Substrate conversion efficiency is highest at the lowest substrate concentrations. The reduction in conversion efficiency at higher substrate concentrations indicates that the substrate is no longer fully consumed at the dilution rate, leading to maximum productivity. According to the simulation results, there is an optimal substrate concentration in the feed that generates both the highest fermenter productivity and substrate conversion efficiency.

When continuous fermentation is carried out under these optimal conditions, at steady state, the substrate is completely consumed and fermentation reaches biomass and metabolite concentrations that correspond to maximum fermenter productivity (Figures 2.13(e) and (f)). When representing fermentation dynamics, it can be worth standardizing fermentation time by the residence time τ of the medium in the fermenter. The resulting graph of the temporal evolution of concentrations reveals the number of transits required to reach the steady state. According to the simulation results, it generally takes between 5 and 10 residence times to stabilize continuous fermentation. Taking the example of an anaerobic cell developing at a growth rate of 0.05 h^{-1}, the residence time is around 20 h, and the steady state is reached after around 100 h of continuous operation.

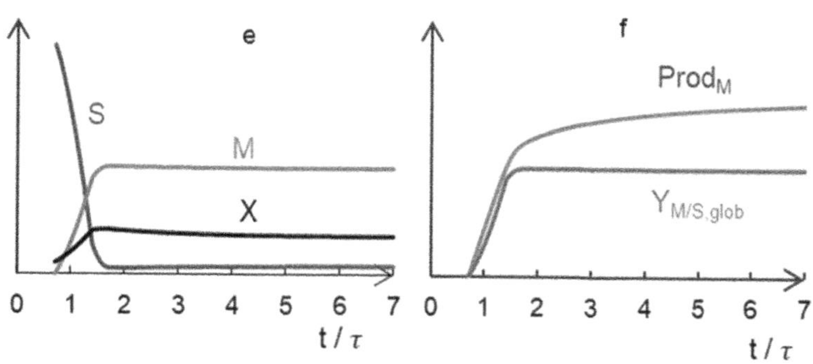

Figure 2.13. *(e) Under optimal operating conditions, variation of substrate S, biomass X and metabolite M concentrations as a function of time standardized to residence time τ. (f) Variation of productivity $Prod_M$ and global conversion efficiency $Y_{M/S,glob}$ as a function of time standardized to residence time τ. For a color version of this figure, see www.iste.co.uk/ghoul/bioreactor.zip*

2.3.7. *Optimization of continuous aerobic fermentation*

2.3.7.1. *Maximum cell concentration and productivity allowed by oxygen transfer*

As previously explained for batch fermentations, in the case of a continuous fermentation involving an aerobic microbial cell, the achievable cell concentration is limited by the transfer of oxygen between the air and the medium. At the steady state, the maximum biomass concentration X_{max} is determined from the maximum oxygen consumption rate by the following relationship:

$$X_{max} = \frac{k_l a\,(O^*_{2,mean} - O_{2l,min})}{Y_{O2/X}\,D} \qquad [2.60]$$

In this relationship, the specific growth rate is replaced by the dilution ratio of the continuous fermenter. The relationship indicates that the maximum biomass concentration decreases with increasing dilution ratio.

The corresponding maximum biomass productivity, $Prod_{X,max}$, can be calculated by the following relationship:

$$Prod_{X,max} = DX_{ss} = \frac{k_l a\,(O^*_{2,m} - O_{2l,min})}{Y_{O2/X}} \qquad [2.61]$$

This is equal to the maximum oxygen transfer rate divided by the stoichiometric coefficient $Y_{O2/X}$. It is important to note that the maximum productivity does not depend on the dilution rate when the average oxygen solubility in the medium can be assumed to be constant (Figure 2.14).

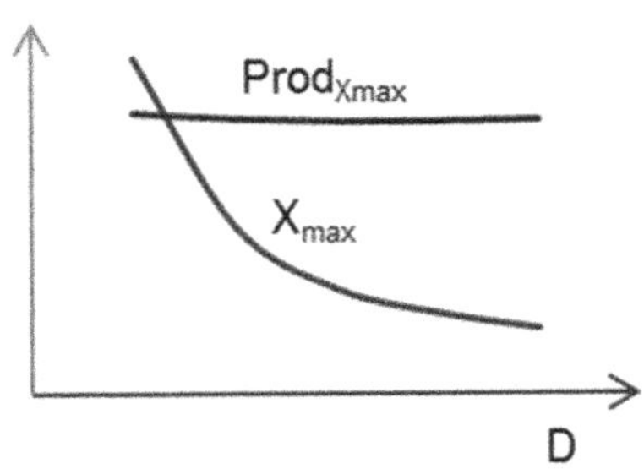

Figure 2.14. *For aerobic cell production in continuous fermentation, influence of dilution rate D on maximum biomass concentration X_{max} and maximum productivity $Prod_{Xmax}$ allowed by oxygen transfer*

2.3.7.2. *Optimization of continuous aerobic fermentation for cellular biomass production*

A second example illustrating the value of continuous fermentation is the aerobic production of microbial cells, accompanied by the simultaneous excretion of undesirable metabolites such as ethanol, acetate or lactate, when the metabolic flux in glycolysis pathways reaches excessive levels. By carrying out fermentation at low carbon substrate concentrations, it becomes possible to considerably reduce the production of these undesirable metabolites.

Simulation of aerobic microbial cell production in a continuous fermenter makes it possible to predict variations in substrate, biomass and metabolite concentrations when the steady state is reached (Figure 2.15(a)). According to the results of this simulation, below a critical dilution rate, the stationary metabolite concentration becomes very low. This observation is explained by the considerable reduction in metabolite excretion when cells are stabilized at a low substrate concentration in the medium. Under such conditions, more substrates become available for biomass production, resulting in an increase in both fermentation productivity and conversion efficiency of substrate to biomass.

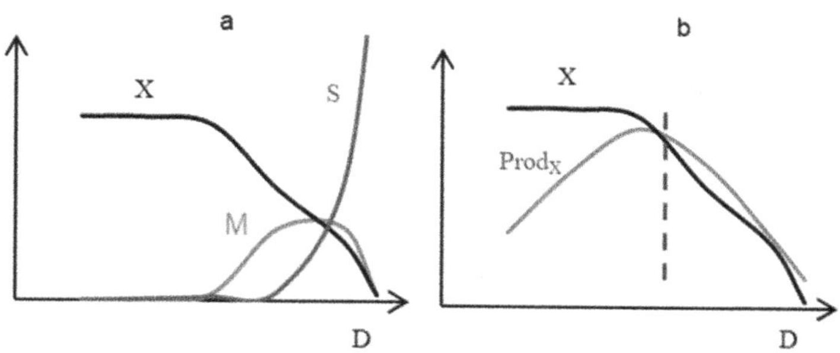

Figure 2.15. *(a) For aerobic fermentation for cell production, influence of dilution rate D on stationary concentrations of substrate S, biomass X and growth-associated metabolite M. (b) Influence of dilution rate D on stationary X cell concentration and fermenter productivity in Prodx biomass. The dotted line indicates the dilution rate for maximum productivity. For a color version of this figure, see www.iste.co.uk/ghoul/bioreactor.zip*

The first step in the optimization procedure involves determining the dilution rate which, at a specified substrate feed concentration, generates the highest fermenter productivity (Figure 2.15(b)). According to the simulation results, fermenter productivity, which is equal to the product of media flow rate and stationary cell concentration, peaks at an intermediate dilution rate. This productivity decreases at higher values of D, due to the decrease in cell concentration, and also at lower values of D, due to the reduction in flow rate. Simulation is used to identify the dilution rate for which productivity is highest, as well as to determine how this dilution rate varies with substrate concentration in the feed. For this cell production, the dilution rate corresponding to maximum productivity does not vary significantly with substrate feed concentration.

In the next phase, the simulation assesses the influence of the substrate concentration in the feed medium (Figure 2.15(c)). An increase in substrate concentration leads to an increase in the stationary biomass concentration, until the maximum cell concentration for oxygen transfer is reached. At the same time, fermenter productivity increases and reaches maximum productivity, while substrate conversion efficiency remains close to its highest value (Figure 2.15(d)). As a result, the optimal substrate concentration in the feed is that which achieves the highest cell concentration, with the aeration capacity of the fermenter in mind.

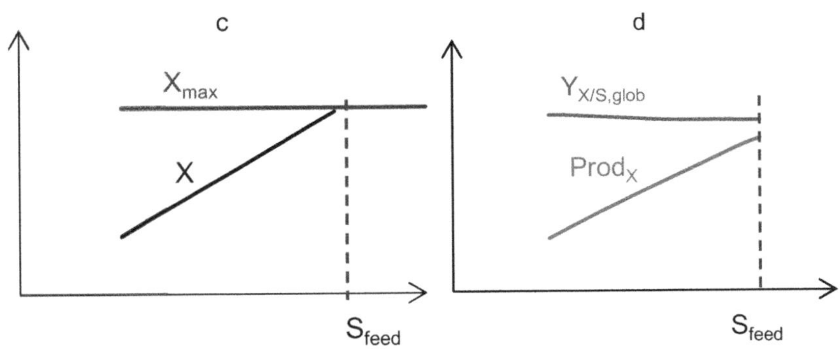

Figure 2.15. *(c) Influence of substrate concentration in feed S_{feed} on stationary concentration X and maximum cell concentration X_{max} allowed by oxygen transfer. The dotted line indicates the S_{feed} concentration for which the stationary concentration is the maximum concentration allowed by aeration. (d) Influence of substrate concentration in feed S_{feed} on $Prod_X$ fermenter productivity and global conversion efficiency of substrate to biomass $Y_{X/S,glob}$. The dotted line indicates the S_{feed} concentration for which productivity is at its maximum. For a color version of this figure, see www.iste.co.uk/ghoul/bioreactor.zip*

When continuous fermentation is run under these optimal conditions, at steady state, the oxygen consumption rate is close to the maximum oxygen transfer rate (Figure 2.15(e)). Residual substrate is maintained at a very low level. Furthermore, the stationary metabolite concentration is negligible and the biomass concentration reaches a level that corresponds to maximum fermenter productivity (Figure 2.15(f)).

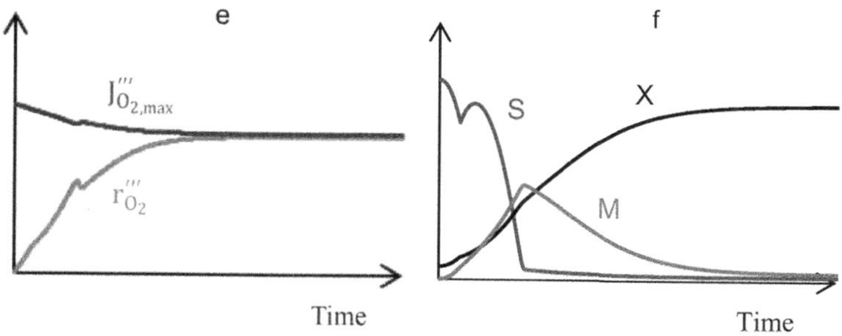

Figure 2.15. *(e) Under optimal conditions for continuous fermentation, variation of the volumetric rate of oxygen consumption r'''_{O2} and the maximum rate of oxygen transfer $J'''_{O2,max}$ as a function of time. (f) Under optimal operating conditions, time-dependent variation in concentrations of substrate S, biomass X and metabolite M. For a color version of this figure, see www.iste.co.uk/ghoul/bioreactor.zip*

2.4. Optimal implementation of fed-batch fermentations

2.4.1. *Principle of fed-batch fermentations*

The fed-batch process can be seen as an intermediate mode between batch and continuous fermentation. The fermenter is partially filled with a starting medium before being inoculated with cells. Fed-batch starts with a batch fermentation phase, during which the microbial cells adapt to their new environment and multiply. Subsequently, a continuous flow of feed medium is introduced into the fermenter, without any medium extraction. The feed flow rate is adjusted to maintain a relatively low and constant substrate concentration. Nutrient feeding is generally stopped when the volume of medium reaches the maximum capacity of the fermenter. Fed-batch fermentation ends with a short, final batch phase to consume the remaining substrates.

Fed-batch fermentation offers several advantages. The main one is its ability to be implemented at limiting substrate concentrations. These substrates can include carbon, nitrogen and other nutrient sources. This approach can result in beneficial changes in cellular metabolism, such as reduced excretion of undesirable metabolites or increased production of desired metabolites. Furthermore, as the medium is not continuously extracted, the limitations associated with cell washout that are encountered in continuous fermentation systems are non-existent. Another favorable aspect of fed-batch fermentations lies in their relatively short duration, which helps to reduce the risk of microbial contamination (Minihane and Brown 1986; Hewitt and Nienow 2007).

This being said, fed-batch fermentations do have a number of disadvantages and limitations. The most significant of the latter is the increased complexity of their implementation compared to batch and continuous fermentations, as they involve optimization of a greater number of operational variables. The feed flow rate must be carefully controlled to maintain substrate concentration at a relatively constant level, and this strategy must be adapted to the kinetics of microbial transformation, requiring an additional investment in time for its development and extrapolation. Furthermore, operating the fermenter with an initial partial filling of medium results in a reduction in overall productivity per fermenter unit volume. Increasing the volume of medium also leads to dilution effects on biomass and metabolites, resulting in lower final concentrations. As in the batch mode, fed-batch fermentations are carried out sequentially, with periods of inactivity to empty, clean, sterilize and refill the fermenter. These relatively long idle times also reduce overall fermenter productivity.

Fed-batch fermentations are the most widely used methods for operating fermenters under conditions of limiting substrate concentrations. Fed-batch is used for microbial cell production, to minimize the excretion of undesirable metabolites associated with growth. It is also widely adopted for the production of non-growth-related metabolites, as continuous feeding enables successive optimization of the cell growth and metabolite production phases.

2.4.2. *Operating variables for fed-batch fermentations*

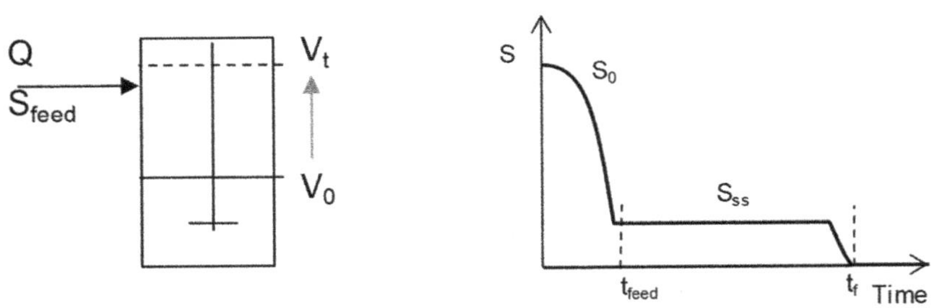

Figure 2.16. *Operating variables in fed-batch fermentation and variation over time in substrate S concentration*

The implementation of fed-batch fermentation involves the optimization of several operational variables associated with the initial batch and continuous medium feed phases (Figure 2.16). For initial batch fermentation, the two main variables are the initial volume fraction V_0/V_t of the medium in the fermenter and the initial composition of the medium, in particular the concentration S_0 of the carbon substrate. Concentrations of other nutrients can also be considered.

In the second feeding phase, which is carried out at a relatively constant substrate concentration, the main operating variable is this steady-state substrate concentration. It exerts a significant influence on cell metabolism, determining rates of growth, substrate consumption and metabolite production. Medium feed begins at the t_{feed} point, when, during the initial batch period, the substrate concentration reaches the S_{ss} value. The second important variable is the S_{feed} substrate feed concentration. It does not directly affect microbial activities, but does influence the feed flow rate and, consequently, the final biomass and metabolite concentrations. The third operational variable to consider is the final time for media feeding. In general, this corresponds to the point at which the internal volume of the medium reaches the total volume of the fermenter. In the case of aerobic fermentations, the final feeding time may also coincide with the moment when the dissolved oxygen concentration falls below the critical oxygen limitation threshold.

2.4.3. *Simulation model for fed-batch fermentation*

Compared to previously established models for batch and continuous fermentations, the simulation models used to optimize fed-batch fermentations are based on the same laws of microbial kinetics, oxygen solubility and oxygen transfer rate. However, these models are distinguished by modified instantaneous balances, which take into account the continuous feeding of medium and the variation in medium volume during fermentation.

2.4.3.1. *Variation in medium volume during fermentation*

The volume of medium in the fermenter increases during fed-batch fermentation, simultaneously with the addition of medium. The temporal variation in the mass of medium m_f in the fermenter is described by the instantaneous total material balance of the fermenter, as a function of the feed flow rate Q and the density ρ_{feed} of the feed medium:

$$Q\rho_{feed} = \frac{dm_f}{dt} \qquad [2.62]$$

The mass of medium m_f present in the fermenter is equal to the product of the volume V of medium and its density ρ_f. Assuming that the density of the medium remains constant, the total material balance equation becomes:

$$Q\rho_{feed} = \rho_f \frac{dV}{dt} \qquad [2.63]$$

Assuming similar values for ρ_{feed} and ρ_f, the time variation in media volume is simply equal to the media feed flow rate:

$$\frac{dV}{dt} = Q \qquad [2.64]$$

Instead of volume V, we usually use the volume fraction V/V_t, in which volume V is divided by the total useful volume V_t of the fermenter. The variation over time of the volume fraction V/V_t is then given by the following relationship:

$$d(\frac{V}{V_t})/dt = \frac{Q}{V_t} \qquad [2.65]$$

2.4.3.2. *Biomass, substrate and metabolite balance for fed-batch fermentation*

In fed-batch mode, the instantaneous cell biomass balance is expressed by the following relationship:

$$r_X = \frac{dm_X}{dt} \qquad [2.66]$$

Since volume V and cell concentration X are both variable, the balance becomes:

$$r_X''' V = \frac{d(V\,X)}{dt} = V\frac{dX}{dt} + X\frac{dV}{dt} \qquad [2.67]$$

and replacing dV/dt by Q, the balance is written as:

$$r_X''' = \frac{dX}{dt} + \frac{Q}{V}X \qquad [2.68]$$

The temporal variation of biomass concentration is thus described by the following relationship:

$$\frac{dX}{dt} = \mu X - \frac{Q}{V}X \qquad [2.69]$$

Similarly, for a metabolite excreted during fermentation, the variation in its concentration is described by the following relationship:

$$\frac{dM}{dt} = q_M X - \frac{Q}{V}M \qquad [2.70]$$

For fed-batch mode, the instantaneous substrate balance is:

$$Q\,S_{feed} - r'''_S V = \frac{d(VS)}{dt} = V\frac{dS}{dt} + S\frac{dV}{dt} \qquad [2.71]$$

or

$$Q\,S_{feed} - r'''_S V = V\frac{dS}{dt} + QS \qquad [2.72]$$

The time variation of the substrate concentration is thus given by the following relationship:

$$\frac{dS}{dt} = -q_S X + \frac{Q}{V}\,(S_{feed} - S) \qquad [2.73]$$

Interestingly, the instantaneous biomass, substrate and metabolite balances in fed-batch mode are identical to those in continuous fermentation. The only distinction lies in the volume of medium V, which is constant in the case of continuous fermentation, whereas it varies during fed-batch fermentation. As a result, fed-batch fermentation can be regarded as a continuous fermentation in which the continuous removal of medium is replaced by an accumulation of medium inside the fermenter.

2.4.3.3. *Variation in cell mass during stationary phase*

At a constant substrate concentration, the specific growth rate of the cells can be considered constant, as long as the concentration of excreted metabolites has no inhibiting influence and oxygen transfer is not limiting. Under these conditions, the biomass balance is as follows:

$$r'''_X V = \frac{d(V\,X)}{dt} \qquad [2.74]$$

or

$$\mu\,XV = \frac{d(V\,X)}{dt} \qquad [2.75]$$

The balance leads to the differential equation for the variation in cell mass VX in the fermenter:

$$\mu dt = \frac{d\,(VX)}{VX} \qquad [2.76]$$

The differential equation is integrated as follows:

$$\mu \int_0^t dt = \int_{X_0}^{X} \frac{dVX}{VX} \qquad [2.77]$$

$$\mu t = \ln \frac{VX}{V_0 X_0} \qquad [2.78]$$

or

$$VX = V_0 X_0 e^{\mu t} \qquad [2.79]$$

According to this equation, when the cell growth rate is kept constant during the substrate feed phase, the cell mass in the fermenter increases exponentially.

2.4.3.4. *Medium feed flow rate*

The steady-state substrate balance can be used to assess the substrate feed flow rate. At steady state, the derivative dS/dt is set to 0 and the substrate balance becomes:

$$dS = 0 = -q_S X + \frac{Q}{V}(S_{feed} - S_{ss}) \qquad [2.80]$$

or

$$\frac{Q}{V}(S_{feed} - S_{es}) = q_S X \qquad [2.81]$$

Thus, the substrate feed flow rate required to maintain a constant substrate concentration is given by:

$$Q = \frac{q_S X V}{S_{feed} - S_{ss}} \qquad [2.82]$$

According to this relationship, the media feed flow rate must be instantly adjusted to the total substrate consumption rate in the fermenter. As the cell mass increases exponentially during fermentation, the media feed flow rate must also increase exponentially.

The feed flow rate can also be expressed as a function of the volume fraction of the medium V/V_t by the following relationship:

$$Q/V_t = \frac{\frac{q_S\,X\,V}{V_t}}{S_{feed} - S_{es}} \qquad [2.83]$$

2.4.3.5. *Oxygen balances for fed-batch fermentation*

The variation in medium volume during fed-batch alters the instantaneous oxygen balance over the entire fermenter, since oxygen consumption only occurs in the medium volume inside the fermenter. The balance is written as:

$$32\,\frac{Q_{a,input}}{V_{m,input}}\,y_{O2,input} - r'''_{O2}V = 32\,\frac{Q_{a,output}}{V_{m,output}}\,y_{O2,output} \qquad [2.84]$$

When the input air flow rate $Q_{a,input}$ is kept constant during fermentation, the mole fraction $y_{O2,output}$ of oxygen in the output air depends on the medium fraction V/V_t; this is calculated by the following relationship:

$$y_{O2,output} = \frac{Q_{a,input}\,V_{m,output}}{V_{m,input}\,Q_{a,output}}\,y_{O2,input} - \frac{\frac{r'''_{O2}\,V}{V_t}}{32\,\frac{Q_{a,output}}{V_t\,V_{m,output}}} \qquad [2.85]$$

The second instantaneous oxygen balance on the medium phase, which expresses the equality between oxygen transfer and consumption rates, remains unchanged, since both rates are proportional to the medium volume. The oxygen balance, expressed per liter of medium, is written as:

$$k_la\left(O^*_{2,m} - O_{2l}\right) = r'''_{O2} \qquad [2.86]$$

2.4.3.6. *Average oxygen solubility in the fermenter*

The variation in medium volume must also be considered when calculating the average oxygen solubility in the fermenter, as input air pressure varies with medium volume. At a constant output air pressure P_{out}, the input air pressure P_{in} depends on the height h_m of the medium in the fermenter, according to the following relationship:

$$P_{input} = P_{output} + \rho_f\,g\,h_m \qquad [2.87]$$

If h_t is the total height of the medium when the fermenter is filled, P_{in} can be expressed as a function of the fraction of medium volume V/V_t by the relationship:

$$P_{input} = P_{output} + \rho g\, h_t \frac{V}{V_t} \qquad [2.88]$$

This variable input pressure is used to calculate the average partial pressure $P_{O2,m}$ and the average solubility $O^*_{2,m}$ of oxygen in the fermenter.

2.4.3.7. *Maximum oxygen transfer rate*

From the average oxygen solubility in the fermenter, the maximum oxygen transfer rate $J'''_{O2,max}$ is determined by the following relationship:

$$J'''_{O2,max} = k_l a\left(O^*_{2,m} - O_{2l,min}\right) \qquad [2.89]$$

When air flow and volumetric fermenter agitation power are kept constant during fermentation, the oxygen transfer coefficient $k_l a$ can be considered constant, and the maximum biomass concentration allowed by the oxygen transfer capacity is assessed by the relationship:

$$X_{max} = \frac{J'''_{O2,max}}{q_{O2}} = \frac{k_l a\,(O^*_{2,m} - O_{2l,min})}{Y_{\frac{O2}{X}}\, \mu} \qquad [2.90]$$

This maximum concentration is inversely proportional to the specific rate of cell growth during fermentation.

2.4.4. *Substrate productivity and conversion efficiency for fed-batch fermentation*

The simulation model also predicts temporal variations in fermenter productivity and the conversion efficiency of substrate to product during fed-batch fermentation. If the product is microbial biomass, $Prod_X$ productivity is calculated as the mass of cellular biomass produced from the start of fermentation, divided by the total fermenter volume V_t and by the total fermentation time, including the additional time between two fermentations:

$$Prod_X = \frac{V X - V_0 X_0}{V_t\,(t + t_{ad})} \qquad [2.91]$$

If the product is a metabolite, $Prod_M$ productivity is calculated as the ratio of the mass of metabolite produced, divided by the total volume of the fermenter and the total duration of fermentation:

$$Prod_M = \frac{V\,M}{V_t\,(t + t_{ad})} \quad [2.92]$$

With regard to the global conversion efficiency of the substrate into product, if the product is microbial biomass, the conversion efficiency $Y_{X/S,glob}$ is calculated as the ratio between the mass of biomass produced, divided by the total mass of substrate added into the fermenter:

$$Y_{M/S,glob} = \frac{V\,X - V_0\,X_0}{V_0\,S_0 + m_{S,add}} \quad [2.93]$$

The mass of substrate added during feeding is determined by integrating the substrate input rate:

$$m_{S,add} = \int_0^t Q\,S_{feed}\,dt \quad [2.94]$$

If the product is a metabolite, the conversion efficiency $Y_{M/S,glob}$ is calculated as the ratio between the mass of metabolite produced, divided by the total mass of substrate used:

$$Y_{M/S,glob} = \frac{V\,M}{V_0\,S_0 + m_{S,add}} \quad [2.95]$$

2.4.5. *Optimization of fed-batch fermentation for cell production*

2.4.5.1. *Optimization objective*

The use of a simulation model to optimize the implementation of fed-batch fermentation is illustrated by an initial case study, involving the aerobic production of bacterial or yeast cell biomass. Cellular growth may be accompanied by the simultaneous production of a metabolite, such as ethanol or acetate, whose excretion results from excessive flux in the glycolytic pathway. By carrying out fermentation at a low carbon substrate concentration, it becomes possible to significantly reduce metabolite excretion, leading to increased conversion of substrate to biomass.

To optimize fermentation, it is necessary to know the kinetics of microbial transformation, in particular the dependence of growth and metabolism rates on the concentration of the carbon substrate, which, in this case, is a sugar (Figures 2.17(a) and (b)). For the microbial transformation considered, the specific growth rate follows a Monod law, while the specific rate of metabolite excretion exhibits a sigmoid-type dependence on sugar concentration. It is characterized by a threshold sugar concentration below which metabolite excretion is negligible. To determine the feed flow rate and dissolved oxygen concentration, we also need to know how the specific rates of sugar and oxygen consumption vary as a function of the sugar concentration in the medium.

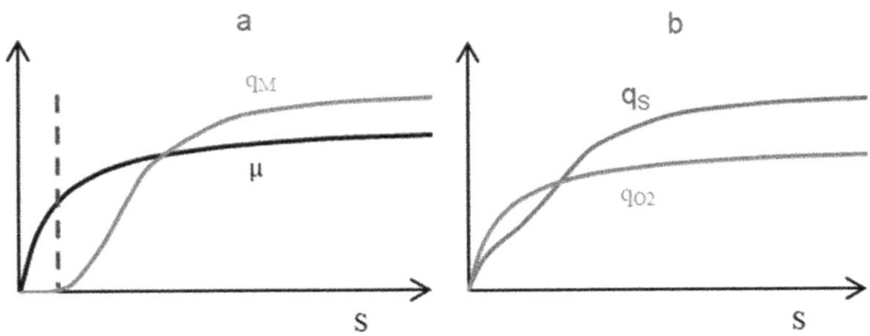

Figure 2.17. *(a) For a production of cells, variation in specific growth rate μ and metabolite production rate q_M as a function of carbon substrate S concentration. (b) Variation in specific substrate consumption rates q_s and oxygen consumption rates q_{O2} as a function of substrate S concentration. For a color version of this figure, see www.iste.co.uk/ghoul/bioreactor.zip*

For this fermentation, four operational variables need to be optimized: the initial volume fraction of the medium, the initial substrate concentration, the substrate feed concentration and the stationary substrate concentration. Fermentation is stopped either when the fermenter is full, or when the oxygen level drops below the minimum level. The optimization procedure aims to achieve both the highest fermenter productivity and substrate-to-biomass conversion efficiency.

2.4.5.2. *Methodology for optimizing operating variables*

The four operating variables in fed-batch fermentation are not independent of one another in terms of their impact on fermentation progress. Indeed, the influence of one variable may depend on the values specified for the other three. Simulation is used to understand the influence of operating variables on the process of fermentation, and then to assess their optimal values. The recommended optimization method consists of optimizing the variables in the following order:

the stationary substrate concentration S_{ss}, the initial substrate concentration S_0, the substrate concentration in feed S_{feed} and the initial medium volume fraction V_0/V_t.

2.4.5.2.1. Stationary substrate concentration

The first step is to identify the optimal range of stationary substrate concentration during feeding. The aim is to achieve the highest specific cell growth rate, while at the same time ensuring negligible metabolite excretion. The influence of stationary substrate concentration is further assessed by simulating fed-batch fermentation at increasing S_{ss} stationary concentrations (Figure 2.18(a)). As observed, the final metabolite concentration increases with S_{ss}. It becomes negligible below an intermediate value of S_{ss} when the rate of metabolite excretion is greatly reduced. The dependence of final biomass concentration on S_{ss} is more complex. It first increases with S_{ss}, reaches a maximum value and then gradually decreases. The observed increase is due to the increase in a specific growth rate. At low S_{ss} values, medium feeding is interrupted when filling the fermenter. At high S_{ss} values, fermentation is halted by oxygen limitation, and the observed decrease in cell concentration is a result of the reduction in maximum biomass concentration allowed by oxygen transfer as growth rate increases.

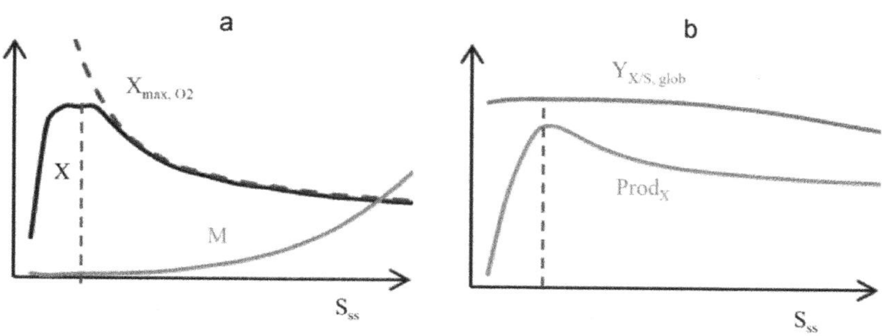

Figure 2.18. *(a) Variation, as a function of the stationary substrate concentration S_{ss}, of the biomass concentration X, the metabolite concentration M and the maximum biomass concentration X_{max} allowed by oxygen transfer. (b) Variation, as a function of the stationary substrate concentration S_{ss}, of fermenter biomass productivity $Prod_X$ and global conversion efficiency of substrate to biomass $Y_{X/S,glob}$. The dotted line indicates the S_{ss} concentration at which fermenter productivity is at its maximum. For a color version of this figure, see www.iste.co.uk/ghoul/bioreactor.zip*

When the influence of the stationary substrate concentration on fermenter productivity is determined, maximum productivity is achieved at the intermediate S_{ss} concentration, which leads to the highest cell concentration (Figure 2.18(b)). This intermediate S_{ss} concentration of maximum productivity is therefore selected. It also ensures a negligible metabolite concentration and thus a maximum substrate-to-biomass conversion efficiency. This optimal S_{ss} concentration also corresponds to the threshold substrate concentration above which metabolite excretion becomes significant.

2.4.5.2.2. Initial substrate concentration

The initial substrate concentration S_0 has little influence on the kinetics of fed-batch fermentation, since the initial batch process takes place over a very short period. It is chosen slightly above the optimal S_{ss} concentration, so that the S_{es} concentration is reached after a few hours of fermentation.

2.4.5.2.3. Substrate concentration in feed

The substrate concentration in the feed S_{feed} does not directly influence cell kinetics. It determines the volumetric flow rate of the medium and the rate of change of the medium volume. It also influences the cell concentration that can be achieved. Simulation is used to more accurately assess the influence of S_{feed} both on fermenter productivity at the previously determined stationary concentration and on global conversion efficiency of substrate to biomass (Figure 2.18(c)).

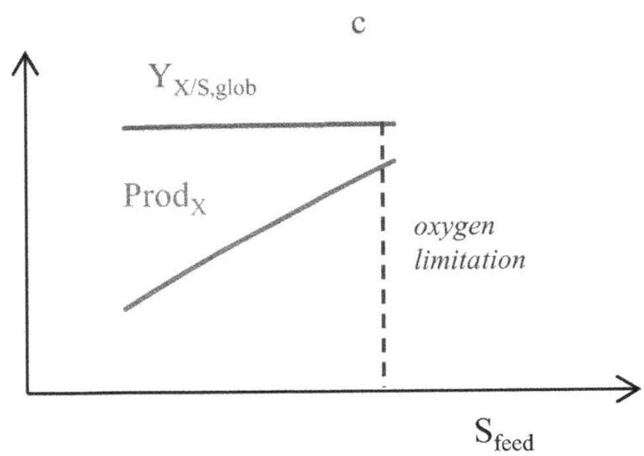

Figure 2.18. *(c) Influence of substrate concentration in feed S_{feed} on fermenter productivity Prodx and overall substrate conversion efficiency $Y_{X/S,glob}$. The dotted line indicates the S_{feed} concentration at which productivity reaches the maximum value allowed by oxygen transfer. For a color version of this figure, see www.iste.co.uk/ghoul/bioreactor.zip*

This figure shows that productivity increases with S_{feed} up to a maximum value corresponding to the maximum biomass concentration allowed by oxygen transfer. At the same time, substrate conversion remains at its maximum. Consequently, the optimal S_{feed} concentration is that which leads to maximum cell concentration and productivity.

2.4.5.2.4. Initial volume fraction of medium

At specified feed and stationary substrate concentrations, the initial volume fraction of the medium V_0/V_t determines when the fermenter is filled and when feeding is interrupted. Simulation of the influence of V_0/V_t at the previously established optimal substrate concentrations reveals an intermediate initial fraction at which fermenter productivity is maximized (Figure 2.18(d)). At lower values of V_0/V_t, the maximum productivity allowed by oxygen transfer is reached before the fermenter is filled. On the other hand, at higher values of V_0/V_t, the fermenter fills up before the maximum cell concentration is reached through aeration. The substrate-to-biomass conversion efficiency remains essentially constant at all V_0/V_t values. Consequently, the optimal initial fraction of medium volume is that which leads to the highest fermenter productivity. At this value, the maximum concentration of cells allowed by the aeration capacity of the fermenter is reached when the fermenter is filled.

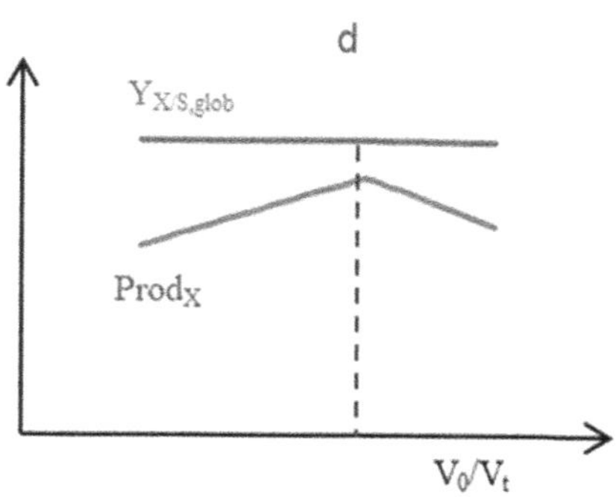

Figure 2.18. *(d) Influence of initial medium volume fraction V_0/V_t on fermenter productivity $Prod_X$ and overall substrate conversion efficiency $Y_{X/S,glob}$. The dotted line indicates the V_0/V_t fraction for which productivity is at its maximum. For a color version of this figure, see www.iste.co.uk/ghoul/bioreactor.zip*

2.4.5.3. *Optimized fed-batch fermentation process*

On the basis of the previously defined operating conditions, the simulation can predict the course of fed-batch fermentation for cell production under optimal conditions (Figures 2.19 (a)–(d)). Fermentation begins with a short batch phase. When the substrate concentration reaches the specified stationary value, media feeding is started and controlled, so as to keep the substrate concentration constant.

During the feed period, there is an exponential increase in cell concentration, negligible metabolite excretion and an exponential increase in feed flow rate. Feeding is stopped when the fermenter is filled, coinciding with the moment when biomass concentration reaches the maximum value allowed by the fermenter's aeration capacity. During fed-batch, the oxygen consumption rate increases until it reaches the maximum oxygen transfer rate. The oxygen concentration simultaneously drops to its minimum critical value. Fermentation ends with a very short additional period to consume the residual substrate remaining in the medium.

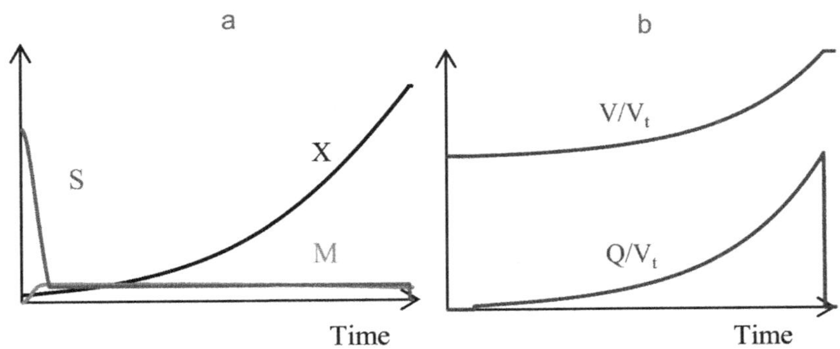

Figure 2.19. *(a) Under optimal fed-batch fermentation conditions, variation over time of substrate S, biomass X and metabolite M concentrations. (b) Under optimal fed-batch fermentation conditions, variation over time of the feed flow rate Q, standardized to the total fermenter volume V_t and of the fraction V/V_t of medium volume in the fermenter. For a color version of this figure, see www.iste.co.uk/ghoul/bioreactor.zip*

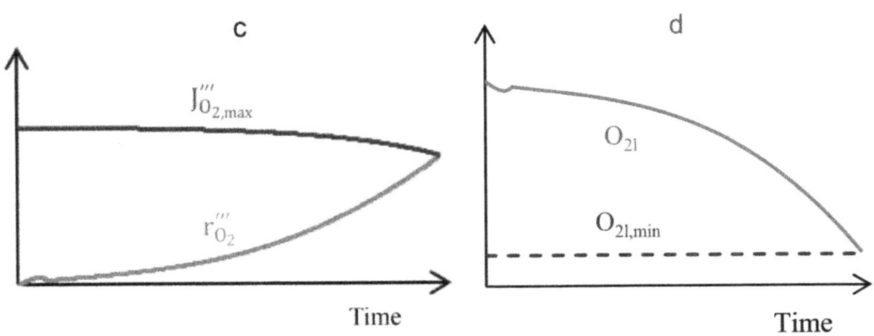

Figure 2.19. *(c) Under optimal fed-batch fermentation conditions, variation over time of the volumetric rates of oxygen consumption r'''_{O2} and maximum oxygen transfer $J'''_{O2,max}$. (d) Under optimum fed-batch fermentation conditions, variation over time of dissolved oxygen concentration O_{2l} and minimum oxygen concentration $O_{2l,min}$*

2.4.6. *Optimization of fed-batch fermentation for metabolite production*

2.4.6.1. *Optimization objective*

The second case study considered is the fed-batch production of a non-growth associated metabolite. Due to the phenomenon of catabolite repression, metabolite synthesis only occurs at a reduced growth rate. Optimization of fed-batch implementation takes the kinetics of microbial transformation into account, in particular the dependence of growth and metabolism rates on the concentration of a carbon substrate, usually a sugar (Figures 2.20 (a)–(c)).

For the microbial cell under consideration, the growth rate follows a Monod law, while the metabolite production rate is inhibited at high substrate concentrations due to catabolite repression. Both cell growth and metabolite production rates are inhibited at high metabolite concentrations. Substrate and oxygen consumption rates are associated with both growth and metabolite production rates.

Fermentation aimed at producing this metabolite is more complex to optimize than the previously described cell production fermentation. It comprises two successive phases: first, a rapid cell growth phase, followed by a metabolite excretion phase under limiting substrate conditions. The aim is to optimize these two phases simultaneously in order to achieve maximum fermenter productivity and the highest conversion efficiency of substrate to metabolite.

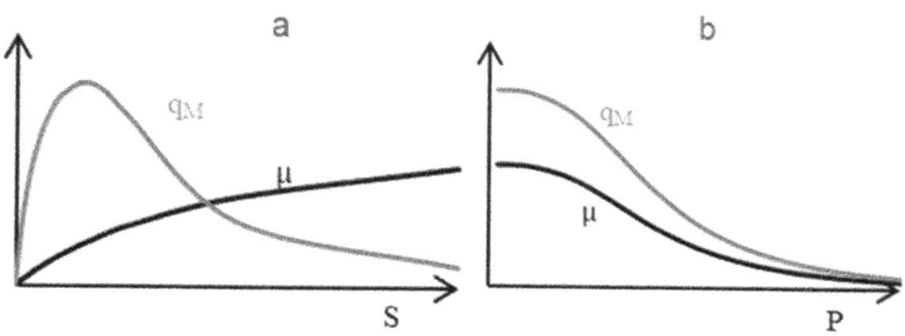

Figure 2.20. *(a) For aerobic production of a non-growth associated metabolite, variation in specific growth rate μ and metabolite production rate q_M as a function of carbon substrate S concentration. (b) Variation in specific growth rate μ and metabolite production rate q_M as a function of metabolite M concentration. For a color version of this figure, see www.iste.co.uk/ghoul/bioreactor.zip*

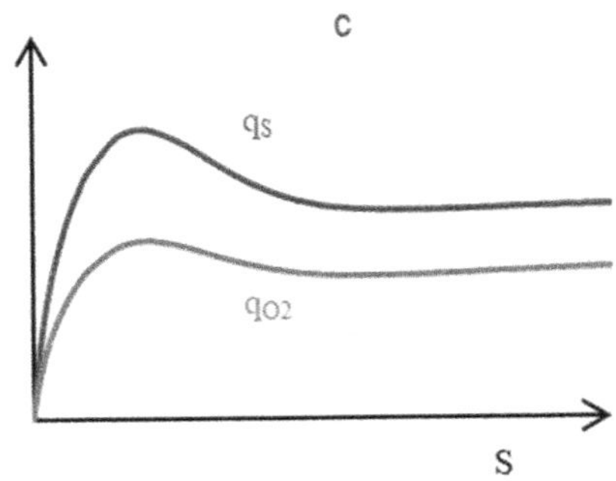

Figure 2.20. *(c) Variation of specific substrate consumption rates q_s and oxygen consumption rates q_{O2} as a function of substrate S concentration*

2.4.6.2. *Methodology for optimizing operating variables*

As in the previous study, simulation is used to understand the influence of operating variables on the process of fermentation and to assess their optimal values. For this metabolite production, the suggested method optimizes the variables in the following order: initial substrate concentration S_0, stationary substrate concentration S_{ss}, substrate concentration in feed S_{feed} and initial medium volume fraction V_0/V_t.

2.4.6.2.1. Initial substrate concentration

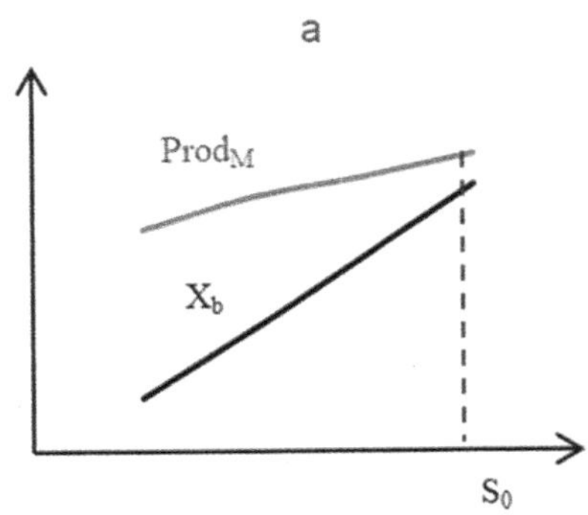

Figure 2.21. *(a) Influence of initial substrate concentration S_0 on biomass concentration X_b at the end of the initial batch and metabolite productivity in $Prod_M$ fed-batch fermentation. The dotted line indicates the maximum cell concentration allowed by oxygen transfer. For a color version of this figure, see www.iste.co.uk/ghoul/bioreactor.zip*

To optimize this metabolite fermentation, the first operational variable to consider is the initial substrate concentration S_0. The aim of the initial batch is to achieve the fastest cell growth and the highest biomass concentration. As the simulation shows, increasing the initial sugar concentration leads to an increase in cell production and metabolite productivity in the fermenter (Figure 2.21(a)). However, as in batch and continuous fermentations, the achievable biomass concentration is limited by the oxygen transfer rate. The maximum cell concentration is estimated by dividing the maximum oxygen transfer rate by the specific oxygen consumption rate. The initial substrate concentration S_0 required to reach this biomass concentration can be estimated by multiplying the synthesized biomass by the stoichiometric coefficient $Y_{S/X}$, which characterizes the specific substrate consumption rate.

2.4.6.2.2. Stationary substrate concentration

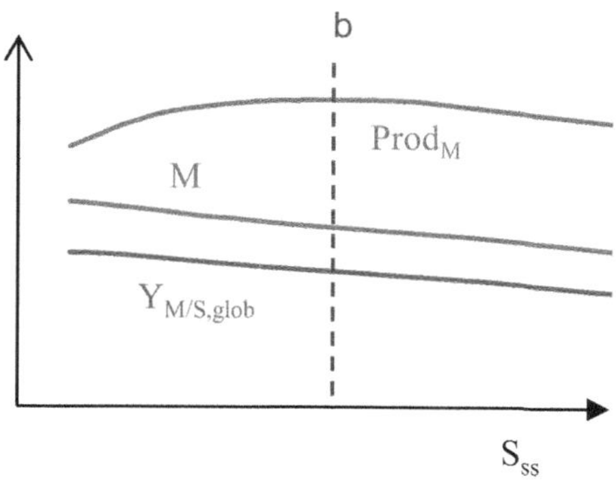

Figure 2.21. *(b) Influence of stationary substrate concentration S_{ss} on final metabolite concentration M, fermenter productivity $Prod_M$ and global conversion efficiency of substrate to metabolite $Y_{M/S,glob}$. The dotted line indicates the S_{ss} concentration at which fermenter productivity is at its maximum. For a color version of this figure, see www.iste.co.uk/ghoul/bioreactor.zip*

The second step is to identify the optimal range of stationary substrate concentration during feeding. The aim is to achieve the highest metabolite production rate during the feeding phase of the fermenter. This is achieved by sufficiently reducing the substrate concentration and the specific growth rate of the cells in order to minimize catabolite repression of metabolite synthesis. The influence of stationary substrate concentration is assessed by simulations of fed-batch fermentation carried out at increasing S_{ss} concentrations (Figure 2.21(b)). As observed, the final metabolite concentration and substrate conversion efficiency decrease with increasing S_s concentration. Productivity peaks at intermediate S_{ss}

concentrations. At lower S_{ss} values, a reduction in productivity is observed due to a lower specific production rate as a result of substrate limitation. At higher S_{ss} values, productivity decreases due to faster specific cell growth and more pronounced repression of metabolite production. Consequently, it is this intermediate concentration of S_{ss}, generating optimal productivity, that is selected. It should be noted that maximum fermenter productivity is not achieved at the highest final metabolite concentration.

2.4.6.2.3. Substrate feed concentration

The third step is to optimize the concentration of S_{feed} substrate in the feed flow. This feed substrate is used by the cells to continue growing at a reduced rate, mainly for metabolite synthesis. S_{feed} concentration does not directly influence cell kinetics. It determines the volumetric flow rate of the medium and the rate of change of the medium volume. It also influences cell and metabolite concentrations. Simulation is used to assess S_{feed}'s influence on fermenter productivity, at the previously determined stationary concentration, final metabolite concentration and the global conversion efficiency of substrate to metabolite (Figure 2.21(c)). It can be seen that metabolite concentration and conversion efficiency increase with S_{feed} concentration. Productivity also increases up to a certain point, and then decreases at higher S_{feed} concentrations. This decrease in productivity is explained by the inhibition of the metabolite production rate at high metabolite concentrations.

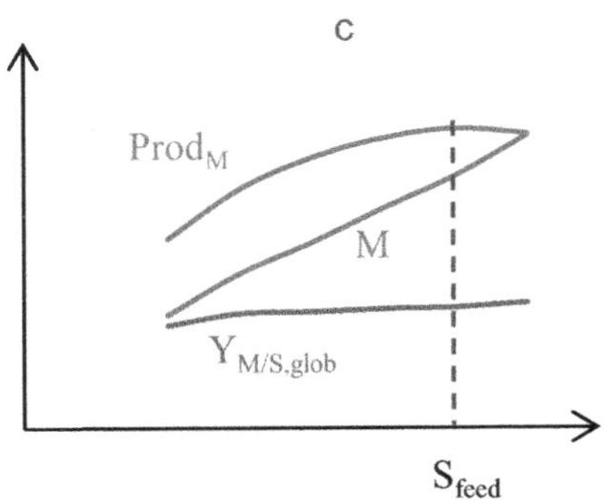

Figure 2.21. *(c) Influence of substrate concentration in S_{feed} feed on final metabolite M concentration, $Prod_M$ fermenter productivity and overall substrate conversion efficiency $Y_{M/S,glob}$. The dotted line indicates the S_{feed} concentration for which fermenter productivity is at its maximum. For a color version of this figure, see www.iste.co.uk/ghoul/bioreactor.zip*

In the context of an optimization strategy that prioritizes fermenter productivity, the optimal S_{feed} concentration chosen is the concentration that leads to maximum metabolite productivity. This substrate concentration is generally relatively high. It also has the advantage of reducing the volume of feed solution to be added to the fermenter, enabling the fed-batch to be started with a higher initial volume of medium, and thus producing a larger cell mass during the initial batch.

2.4.6.2.4. Initial volume fraction of medium

At specified feed and stationary substrate concentrations, the initial volume fraction of the medium V_0/V_t determines when the fermenter is filled and when feeding is stopped. Simulations carried out to assess the impact of V_0/V_t at the previously established optimal substrate concentrations indicate that both final metabolite concentration and global conversion efficiency decrease with V_0/V_t due to the reduction in fermentation time (Figure 2.21(d)). On the other hand, productivity peaks at an intermediate medium volume fraction. At lower values of V_0/V_t, the fermenter is only partially filled when productivity peaks. Conversely, at higher values of V_0/V_t, added volumes of medium and fermentation times are reduced, preventing maximum metabolite productivity from being achieved. As a result, the optimal initial fraction of medium volume is that which leads to the highest fermenter productivity. It is at this value that maximum productivity is reached when the fermenter is filled.

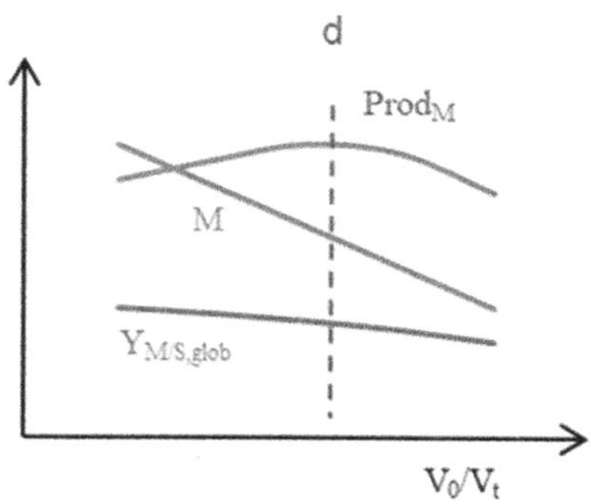

Figure 2.21. *(d) Influence of initial medium volume fraction V_0/V_t on final metabolite concentration M, fermenter productivity $Prod_M$ and overall substrate conversion efficiency $Y_{M/S,glob}$. The dotted line indicates the V_0/V_t fraction for which productivity is at its maximum. For a color version of this figure, see www.iste.co.uk/ghoul/bioreactor.zip*

It should be noted that in this fed-batch fermentation for metabolite production, the operational conditions that maximize productivity do not coincide with those that achieve the highest conversion efficiency of substrate to metabolite. This is because maximum productivity is achieved at relatively high cell biomass concentrations, which requires greater substrate use for biomass production, resulting in lower substrate-to-metabolite conversion efficiency. In the optimization approach described above, fermenter productivity is the main criterion. This approach could be complemented by an economic assessment to determine the impact of both productivity and efficiency criteria on the cost of metabolite production. It is this economic optimization, aimed at reducing the cost of metabolite production, that makes it possible to identify the optimal compromise between fermenter productivity and substrate conversion efficiency.

2.4.6.3. *Optimized fed-batch fermentation process*

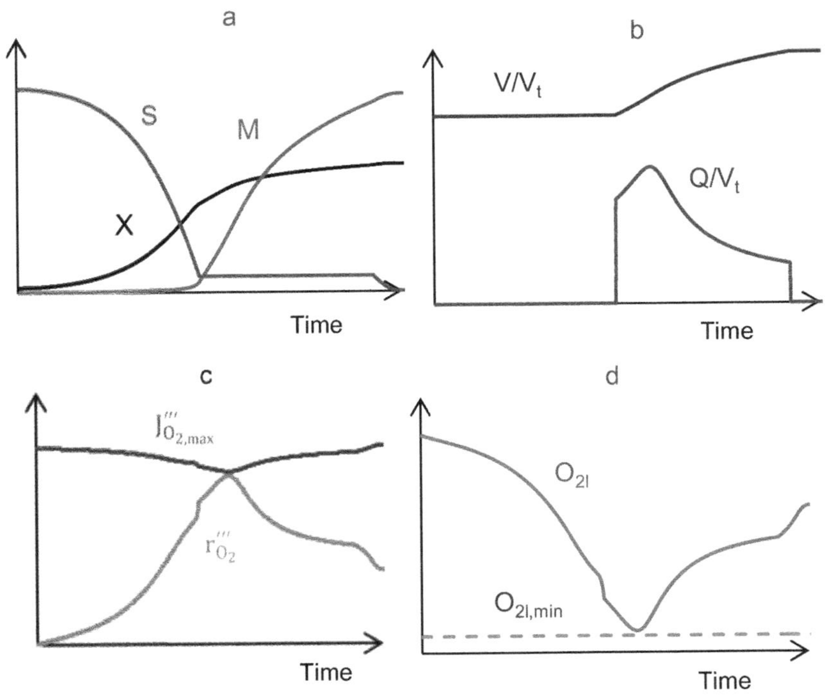

Figure 2.22. *(a) Under optimal fed-batch fermentation conditions, variation over time of substrate S, biomass X and metabolite M concentrations. (b) Under optimal fed-batch fermentation conditions, variation over time of the feed flow rate Q, standardized to the total fermenter volume V_t and of the fraction V/V_t of medium volume in the fermenter. (c) Under optimal fed-batch fermentation conditions, variation over time of the volumetric rates of oxygen consumption r'''_{O2} and maximum oxygen transfer $J'''_{O2,max}$. (d) Under optimal fed-batch fermentation conditions, variation over time of dissolved oxygen concentration O_{2l} and minimum oxygen concentration $O_{2l,min}$. For a color version of this figure, see www.iste.co.uk/ghoul/bioreactor.zip*

Based on the previously defined operating parameters, the simulation predicts the optimal course of fed-batch fermentation for metabolite production (Figure 2.22(a)). Fermentation begins with a batch phase, characterized by rapid cell growth and negligible metabolite production (Figure 2.22(b)). It reaches the maximum biomass concentration made possible by oxygen transfer (Figures 2.22(c) and (d)). When the substrate concentration reaches the optimal stationary value, the feeding of medium starts at a controlled rate to maintain this concentration constant. At this concentration level, metabolite synthesis significantly accelerates, while cell growth continues at a reduced rate. Feeding is interrupted when the fermenter is filled with medium and fermenter productivity simultaneously reaches its maximum value (Figure 2.22(b)).

2.5. References

Atkinson, B. and Mavituna, F. (1991). *Biochemical Engineering and Biotechnology*, 2nd edition. Macmillan, Basingstoke.

Dong, Y., Zhang, Y., Liu, D., Chen, Z. (2023). Strain and process engineering toward continuous industrial fermentation. *Frontiers of Chemical Science and Engineering*, 17, 1–18.

Du, Y.H., Wang, M.Y., Yang, L.H., Tong, L.L., Guo, D.S., Ji, X.J. (2022). Optimization and scale-up of fermentation processes driven by models. *Bioengineering*, 9(9), 473.

Gordeeva, E.L., Ravichev, L.V., Borodkin, A.G., Gordeeva, Y.L. (2021). Mathematical modeling of a biotechnological continuous fermentation process for lactic acid production: A review. *Theoretical Foundations of Chemical Engineering*, 55, 1192–1203.

Hewitt, C.J. and Nienow, A.W. (2007). The scale-up of microbial batch and fed-batch fermentation processes. *Advances in Applied Microbiology*, 62, 105–135.

Koulouris, A., Misailidis, N., Petrides, D. (2021). Applications of process and digital twin models for production simulation and scheduling in the manufacturing of food ingredients and products. *Food and Bioproducts Processing*, 126, 317–333.

Liu, S. (2020). *Bioprocess Engineering: Kinetics, Sustainability, and Reactor Design*, 3rd edition. Elsevier, Amsterdam.

Mears, L., Stocks, S.M., Albaek, M.O., Sin, G., Gernaey, K.V. (2017). Mechanistic fermentation models for process design, monitoring, and control. *Trends in Biotechnology*, 35(10), 914–924.

Minihane, B.J. and Brown, D.E. (1986). Fed-batch culture technology. *Biotechnology Advances*, 4(2), 207–218.

van't Riet, K. and Tramper, J. (1991). *Basic Bioreactor Design*. Marcel Dekker, Boca Raton.

Schügerl, K. and Bellgardt, K.H. (2012). *Bioreaction Engineering: Modeling and Control.* Springer, Berlin, Heidelberg.

Villadsen, J., Nielsen, J., Lidén, G. (2011). *Bioreaction Engineering Principles.* Springer, New York.

Wang, S.J. and Zhong, J.J. (2007). Bioreactor engineering. In *Bioprocessing for Value-Added Products from Renewable Resources.* Yang, S.T. (ed.). Elsevier, Amsterdam.

3

Photobioreactor Engineering

Hatem BEN OUADA[1], Jihène AMMAR[1] and Mohamed GHOUL[2]

[1]*Laboratoire de Biotechnologie Bleu et Bioproduits Aquatiques (B3 Aqua), Institut National des Sciences et Technologies de la Mer (INSTM), Monastir Annex, Tunisia*

[2]*Laboratoire Réaction et Génie des Procédés (LRGP), CNRS, Université de Lorraine, Nancy, France*

This chapter presents an overview of the photobioreactors (PBRs) used in the production of microalgae and associated metabolites for the food industry.

3.1. Overview of different photobioreactor configurations

Microalgae production has a history dating back to the 1960s. Originally, it focused on the use of marine species such as *Tetracelmis* and *Isochrisis*, intended as feed for the early stages of fish and shellfish rearing in hatcheries. Over the years, however, this production evolved into an industry for human and animal consumption, starting in the 1970s.

Initially, production trials often took place in natural ponds without agitation, but as demand grew, more complex culture systems were developed to increase yields and the quality of the microalgae produced. At that time, large-scale production mainly focused on three microalgae species that were particularly tolerant of extreme growing conditions: *Chlorella sp.*, *Arthrospira platensis* and *Dunaliella salina*. These species were favored because of their robustness and ability to thrive in diverse environments (Borowitzka 2013). Over time, however, there has been renewed interest in isolating and identifying new strains of microalgae. This has led to the exploration of new ecosystems and the screening of new strains in order to

Bioreactor Implementation in the Agro-Food Industries,
coordinated by Mohamed GHOUL.

discover new properties and metabolites. This research has opened up new prospects for the microalgae industry, enabling the range of possible products and applications to be diversified.

Production systems have also evolved to meet the changing needs of the industry. They have been adapted to control production factors such as light, temperature, nutrition and the chemical composition of the culture medium, so as to promote the growth of new microalgae species or the synthesis of specific bioactive compounds. Numerous microalgae culture systems have been described in the literature, whose structure, form and degree and manner of control of the various growth factors differ from one system to another.

Overall, microalgae production systems can be classified into two main categories: open systems and closed systems (PBRs). This distinction is based on the level of crop contact with the atmosphere (Apt and Behrens 1999; Borowitzka 2013). Open systems expose a significant part of the culture to the atmosphere. This means that microalgae are grown in the open air, usually in ponds or basins. Growing conditions can vary according to local weather, and this can limit the species that can be grown due to their sensitivity to environmental conditions (Vonshak and Richmond 1988; Blanco et al. 2007; De Andrade et al. 2022).

Closed systems, often known as PBRs, are systems in which microalgae cultures are isolated from the atmosphere. This enables more precise control of the culture environment, including light, temperature, chemical composition of the culture medium and other parameters. PBRs are designed to minimize the risk of contamination and offer greater consistency in terms of product quality.

PBRs offer many advantages, including the ability to maintain continuous cultivation, achieve high volumetric productivity and guarantee consistent product quality (Tang et al. 2012; San Pedro et al. 2014; Sforza et al. 2014; Acién et al. 2019). They are also more flexible in terms of choice of microalgae species and less dependent on local weather conditions. However, designing and operating efficient and cost-effective PBRs remains a challenge. It is necessary to design PBRs with low manufacturing costs, while guaranteeing good light availability, efficient mass transfer and ease of control (Acién et al. 2019; Lindblad et al. 2019).

In the following, we will present an up-to-date overview of the main microalgae cultivation systems in operation.

3.1.1. *Open cultivation systems*

Open cultivation systems (Table 3.1) are the most widely used in industrial operations. They are generally operational outdoors on large areas of land. Their low installation and operating costs (Norsker et al. 2011; Chisti 2013a, 2013b; Costa and de Morais 2014) give them an advantage, but this type of operation is more exposed to contamination and the hazards of environmental factors, which affects productivity and limits the number of species under cultivation (Tredici 2004; Carlsson 2007; Dragone et al. 2010; Vasumathi et al. 2012; De Andrade et al. 2022). Various types of cultivation structure in open systems are described in the literature.

3.1.1.1. *Circular basins*

Circular basins (Figure 3.1) are mainly used in Asia for chlorella cultivation (Singh and Dhar 2011) but are little used for the cultivation of other species. They are often implemented with an agitation system in the form of arms rotating around a central axis or in the form of paddle wheels or turbines with three or four blades (Meng et al. 2015). The depth of the culture varies from a few centimeters to 30 cm, and the size of the basin is limited by the water resistance constraint imposed on the agitation motor (Lee 2001). The largest basin described measures 50 m in diameter (Lee 1997). The use of this cultivation system remains limited, mainly due to the fact that the circular shape does not optimize the operating area and because of the difficulty of extrapolating more than 1,000 m^2 per unit (Lundquist et al. 2010). The productivity of these systems varies, with values ranging from 8.5 to 21 $g/m^2/day$ (Table 3.1).

3.1.1.2. *Raceways*

The raceway configuration (Figure 3.2) represents the most widespread and generalized open system configuration. Indeed, around 90% of the microalgae biomass on the market is produced in raceways (Chisti 2013a). In their basic form, raceways consist of basins built in earth with liner, or in concrete, and with a length/width ratio of around 10–20 (Andersen 2005). A central crosspiece divides the basin in two lengthways. The depth of the culture is no more than 30 cm in order to promote light photosynthesis penetration.

Agitation is essential to prevent sedimentation and optimize light distribution in the culture volume. It is generally provided by paddle wheels, driven by pulley motors with adjustable speed, creating crop movement along the entire length of the basin. A flow speed of between 20 and 30 cm/s is recommended (Andersen 2005).

Figure 3.1. *Circular Chlorella cultivation basins in Japan. Unit area 500 m^2 (source: Lundquist et al. 2010). For a color version of this figure, see www.iste.co.uk/ghoul/bioreactor.zip*

Figure 3.2. *Greenhouse raceways. Unit area of 100 m^2. Spirulina cultivation at Company Eden Life, Kettana, Tunisia. For a color version of this figure, see www.iste.co.uk/ghoul/bioreactor.zip*

Rounded basin ends and/or deflectors placed at the ends reduce pressure drop in the acute bends of the raceway, as well as avoid the presence of concentration or thermal gradients (Pulz and Scheibenbogen 1998). The insertion of sumps along the channel is also common, so as to inject gas such as CO_2 or to remove oxygen (Mendoza et al. 2013). Different sump configurations can be used with or without deflectors and with co-current or counter-current gas-liquid contact (Acién et al. 2013).

In some cases, several basins are interconnected in the form of an algal channel (Figure 3.3) with multiple compartments, separated by a network of deflectors (Sompech et al. 2012). In other cases, the plane of the basin is slightly inclined (1.7% slope), and the culture is collected at the bottom and reinjected at the top of the reactor through a pumping system. This system reduces the thickness of the culture for more efficient access to light, and to facilitate biomass collection and separation (Šetlík et al. 1970; Doucha and Lívanský 1995).

Figure 3.3. *Algal channel under a thermo-regulated, illuminated system. Volume: 250 L. Institut National des Sciences et Technologies de la Mer (INSTM), Monastir, Tunisia. For a color version of this figure, see www.iste.co.uk/ghoul/bioreactor.zip*

Raceway systems can be used to furnish large crop areas ranging from 0.5 to 250 ha (Tredici 2004; Chisti 2013b). They offer the advantages of ease of maintenance, direct exposure to sunlight, self-cooling by evaporation and less oxygen accumulation by releasing it into the atmosphere, in addition to a construction cost that is far lower than that of closed systems (Chisti 2013a). However, cultivation is closely dependent on climatic conditions, with an increased risk of microbial contamination and high CO_2 losses. Today, these systems are reserved for the production of species that can withstand extreme environments, such as halophilic species like *Dunaliella salina* or alkaliphilic species like *Arthrospira platensis*, so as to avoid the proliferation of airborne microorganism contaminants, or "robust" microalgae like *Chlorella*, *Scenedesmus* or *Nannochloropsis*.

This cultivation system is often exposed to the risk of dilution by rainwater, silting or contamination (rotifers, parasites, mosquitoes, etc.). Water evaporation in open basins is also important. Indeed, even if the culture medium is partially recycled after harvesting, the water lost through evaporation should be replaced by new water in order to not affect the salinity and pH of the culture medium. Guieysse et al. (2012) estimate evaporation rates ranging from 0.476 m^3/m^2/year in tropical climates to 2.275 m^3/m^2/year in arid climates. Wigmosta et al. (2011) reported evaporation values of between 4 and 21 million L/ha/year in the western United States, and between 1 and 4 million L/ha/year in the eastern United States.

Raceway systems, and open systems in general, also fail to deliver stable, reproducible production. Indeed, their productivity varies significantly according to seasonal and regional climatic fluctuations. Annual productivities in these systems obtained in temperate zones vary on average from 10 to 20 tons/ha (Benemann and Oswald 1994) to 40 tons/ha (Tredici 2004). Nevertheless, exceptional productivities of 150 tons/ha/year have been reported (Lundquist et al. 2010). In most cases, surface productivities in raceways do not exceed 12 g/m^2/day, even during periods of strong illumination (Table 3.1).

Today, greenhouse raceways are becoming increasingly widespread (Morillas-España et al. 2022). The greenhouse protects the crop from external contamination and ensures relative temperature stability throughout the day, while limiting water loss through evaporation. On some farms, greenhouse productivity has been estimated at between 4 and 13 g/m^2/day (Hase et al. 2000), and even as high as 24.9 ± 0.9 g/m^2/day (Morillas-España et al. 2022). The greenhouse reduces crop cooling during the winter period, but excessive crop heating and high summer insolation can lead to photoinhibition phenomena (Pessi et al. 2022). On some farms, the roof and side walls of the greenhouse are opened above a temperature threshold (35°C for spirulina) and crop shading is activated above a light threshold, thus reducing the inhibitory effect. Total and continuous thermal control of greenhouse crops has been achieved in laboratory structures, but the high energy cost of the operation reduces their large-scale application. Recently, greenhouses equipped with transparent photovoltaic panels have been tested. They would reduce energy costs and ensure high productivity (Bernard 2014).

3.1.1.3. *Bubble columns*

Vertical bubble column PBRs are used outdoors in the open air or in greenhouses (Figure 3.4). They generally comprise transparent vertical columns made of glass or acrylic (Guo et al. 2015; Gongi et al. 2021; Morales-Amador et al. 2021) or fully transparent plastic sheaths (Fuentes et al. 2020). In most cases, the culture is kept in suspension by air or CO_2 injection, mainly at the base of each column. Column diameters range from 10 to 60 cm maximum. Irradiation over the entire height of the reactor column thus ensures greater uniformity in terms of capturing light rays than in raceways, where cell access to light is only limited to the surface. Productivity is greatly improved compared to raceways, with values as high as 0.48 g/L/day in the greenhouse and 0.35 g/L/day (Tredici 2004) to 0.5 g/L/day (Gongi et al. 2021) outside.

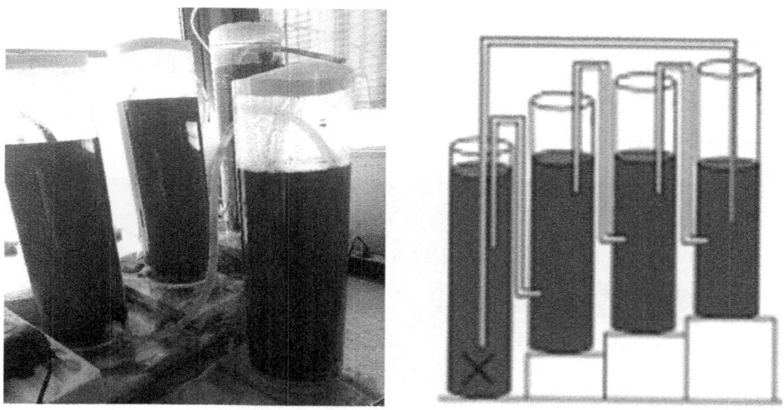

Figure 3.4. *Bubble columns connected in series. Unit volume: 12 L. Institut National des Sciences et Technologies de la Mer (INSTM), Monastir, Tunisia. For a color version of this figure, see www.iste.co.uk/ghoul/bioreactor.zip*

In most cases, the height of the columns is equal to 2.5 m, for a unit culture volume of no more than 400–500 L. Extrapolating these systems to an industrial scale requires the cultivation of a large number of columns. For an equivalent volume, at least 80 columns would be needed to replace a 40 m³ raceway. The individual management of crops in each column is quite complex, even for a small-scale operation.

However, columns can be arranged in series (Stroiazzo-Mougin 2011; Guo et al. 2015; Ben Ali et al. 2022) by polystyrene tubes connecting the columns from bottom to top. A circulation pump submerged in the last column, or in a tray, feeds the first column, allowing the culture to circulate throughout the system. The arrangement of the columns at different levels ensures gravity circulation of the medium. Operating as a single block, the system allows biomass collection and culture management to be located in a single reactor. Zitelli et al. (2013) report on a pilot operation in Spain consisting of six modules of eight columns in series, each measuring 8 m in height.

3.1.1.4. *Thin-film culture*

The main feature of thin-layer cultivation (TLC) systems is the use of shallow cultivation (Figure 3.5) to increase biomass concentration and maximize light utilization efficiency (Acién et al. 2019). The microalgae suspension flows continuously in a very thin layer (often less than 1 cm) over horizontal or sloping platforms (Doucha and Lívanský 2006; Jerez et al. 2014; Morales-Amaral et al. 2015), cascades (Setlík et al. 1970) or sloping raceways (Grivalský et al. 2019). It flows back into a retention basin, where it is pumped through a return pipe to the upper part of the crop, where photosynthesis takes place. The retention basin is used to manage the crop (degassing, supply of CO_2, nutrients, etc.) and harvest the biomass by filtration.

Figure 3.5. *Thin-layer cultivation. Total area 90 m^2. Campus of Institute of Microbiology in Třeboň, Czech Republic (source: Acién et al. 2013). For a color version of this figure, see www.iste.co.uk/ghoul/bioreactor.zip*

It is often protected and designed to store the total volume of the crop, under aeration, during the night and/or periods of non-operation, helping to avoid night frosts and rain, which is one of the advantages of these systems (Doucha and Lívanský 2009).

In some of the configurations described by Becker (1994), the inclination of the cultivation platforms (between 0.5° and 3°) and the speed of the circulation pump are adjustable, enabling the thickness of the culture (between 5 and 15 mm) to be varied according to the intensity of solar irradiation. This is an advantage when it comes to maintaining a favorable light regime in line with the species' physiological requirements and seasonal variations. Another advantage of TLCs is their self-regulation of temperature, as the crop is rapidly heated by solar irradiation in cold periods and cooled by increased evaporation when overheated (Grivalský et al. 2019).

The main advantage of TLCs is that, thanks to their efficient mixing and short optical path, they enable optimal use of light, resulting in very high biomass densities (15–35 g/L; Masojídek et al. 2015) and daily areal productivities of up to 55 g/m^2. High density and low crop volume also minimize the cost of biomass recovery during harvesting. However, the shear forces caused by pumping and flow, as well as wide temperature variations, limit the use of this system to cultivate robust, fast-growing species. Species from the *Scenedesmus* and *Chlorella* genera are the most suitable for cultivation in TLC systems (Torzillo et al. 2010; Grivalský et al. 2019).

Large-scale extrapolation of TLCs is also difficult due to the high surface-to-volume ratios. Compared with raceways with S/V ratios of up to 10 m^2/m^3, the S/V values of TLC systems can reach up to 100 m^2/m^3 (Grivalský et al. 2019). The largest TLC installations listed in the reviews by Acién et al. (2019) and Grivalský et al. (2019) total 900–1,500 m^2 for a cultivation volume of no more than 100 m^3.

3.1.2. *Closed systems: photobioreactors*

In contrast to open systems, closed systems can regulate and control virtually all factors influencing cultivation, enabling production to be controlled for all microalgae species, irrespective of the surrounding climatic conditions. These systems ensure reproducible cultivation with stable productivity, controllable hydrodynamics and constant operating temperature (Pulz 2001), while reducing the risk of contamination, CO_2 losses and water evaporation. Various variants of closed PBRs have been used in the laboratory or on a pilot scale. These systems can have a wide range of shapes, agitation methods, light sources and thermal regulations (Harun et al. 2010; Egbo et al. 2018; Acién et al. 2019).

3.1.2.1. *Tubular photobioreactors*

Only a few types of closed PBR are currently used on a large scale. The most widespread are tubular PBRs (Figure 3.6). They comprise an arrangement of several transparent glass or plastic tubes of varying diameters, lengths and configurations, through which the culture is circulated by pumps or air lift. A mixing tank is often integrated into the system, enabling the removal of oxygen from photosynthesis (degassing) and the control of cultivation parameters (Ugwu et al. 2008; Zittelli et al. 2013; Torzillo and Zittelli 2015; Acién et al. 2019). Tube arrangements can be straight-line or spiral on vertical, horizontal or helical planes (Egbo et al. 2018). The vertical design allows greater mass transfer and lower energy consumption, while the horizontal reactor is more scalable, but requires a large area of land (Ugwu et al. 2008). De Vree et al. (2015) concluded that vertical PBRs generate higher areal productivities than horizontal PBRs due to lower incident photon flux densities at the reactor surface.

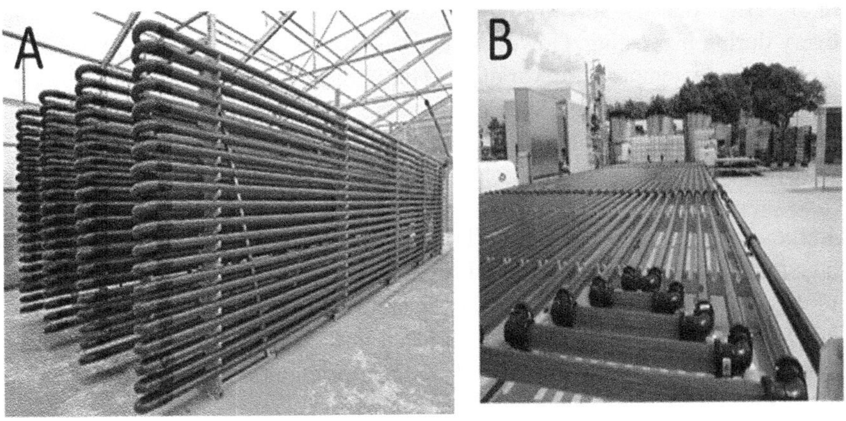

Figure 3.6. *Tubular photobioreactors. (A) Vertical coil. Volume: 4,000 L (source: IGV, Biotech-Ownwork, CC BY-SA 3.0, https://commons.wikimedia.org/w/index.php?curid=25767679). (B) Horizontal. Three loops of 80 m each. Algae PARC pilot facilities, Wageningen UR, the Netherlands (source: De Vree et al. 2015). For a color version of this figure, see www.iste.co.uk/ghoul/bioreactor.zip*

For the same strain (*Scenedesmus obliquus*), areal productivity in tubular PBR can double in comparison with raceway (Arbib et al. 2013). The same finding is observed by De Vree et al. (2015) for the volumetric productivity of *Nannochloropsis*. However, a more detailed analysis of the literature data (Table 3.1) shows strong variations in volumetric productivities that can exceed a factor of ten. These variations are not only due to the strain cultivated, but also to cultivation parameters, unit volume, illumination, production period and, presumably, the frequency of biomass collection and nutrient addition.

Pilot or industrial scale-up of most tubular PBRs described in the literature faces various constraints, notably the high cost of construction materials, operational complexity, biofouling that reduces light accessibility and high energy requirements for mixing and pumping (Abdur Razzak et al. 2013; Liao et al. 2015; Vo et al. 2019).

Tube diameter is a key factor affecting light capture efficiency and, consequently, the biomass productivity of the system. According to Slegers et al. (2013), a diameter of less than 10 cm is required to achieve optimal light efficiency. Brindley et al. (2004) and Torzillo and Zittelli (2015) concluded that a tube diameter between 5 and 9 cm ensures the highest volumetric productivity. For larger tubes, 38 cm in diameter, Huntley and Redalje (2007) observed lower light capture, due to a longer optical path. Note that with 10-cm diameter tubes, it would take over 5 km of tube length to obtain a volume of 40 m^3.

At Roquette Klötze in Germany, microalgae are grown in greenhouses in a tubular reactor with a total volume of 700 m^3, requiring a tube length of 500 km. Compared with raceways of the same volume, the cost of the structure is at least 10 times higher.

Tubes that are too long increase the culture's residence time inside the reactor, causing high oxygen supersaturation, which inhibits photosynthesis (Carvalho et al. 2006). According to Acién et al. (2019), the optimum tube length is 100–150 m, which requires modular reactor units to be assembled instead of using tubes that are too long (Eriksen 2008). As each module is managed individually in terms of circulation, regulation and biomass collection, the cost of the installation becomes extremely high (Kunjapur and Bruce Eldridge 2010).

The operation of tubular PBRs is also affected by the phenomenon of microalgae adsorption to the tube wall, known as "biofouling", especially in aging cultures producing large quantities of exopolysaccharides. Surface deposition reduces light penetration and causes turbulence that impedes efficient use of nutrients. It can lead to crop failure within a few days. Harris et al. (2013) have shown that the use of low-density polyethylene tubes reduces fouling, but even under these conditions, the authors conclude that a cleaning cycle is necessary after a maximum of 12 weeks of cultivation.

3.1.2.2. *Flat-panel PBRs*

Flat-panel PBRs (Figure 3.7) are another variant of closed microalgae culture systems. In their basic version, they consist of flat, rectangular or square panels, usually made of glass, Plexiglas or transparent plastic, positioned horizontally (Camacho-Rodríguez et al. 2014; De Vree et al. 2015) or vertically (Cheng-Wu 2001; Zittelli et al. 2013). The microalgae suspension circulates between the panels in a thin layer of no more than 5–6 cm. The culture is often stirred directly by bubbling air through a perforated tube at the bottom of the reactor (Janssen et al. 2003; Sierra et al. 2008) or by air lift (Kochem et al. 2014; Li et al. 2015).

The main advantage of the flat configuration lies in the relatively short light path through the thin layer of culture and the very large lighting surfaces in relation to the volume (Tredici et al. 1991), thus promoting optimal use of light radiation within the culture. This type of PBR productivity can be quite high, especially for small laboratory structures, reaching 1.7 g/L/day (Zou and Richmond 1999; De Vree et al. 2015) or even up to 2 g/L/day (Tredici et al. 1991). Hu et al. (1998) reported exceptional cell densities of 84 g/L for *Chlorococcum littorale*, grown in the laboratory in a flat PBR with a 1 cm light path and a luminosity of 2,000 µmol photons/m^2/s.

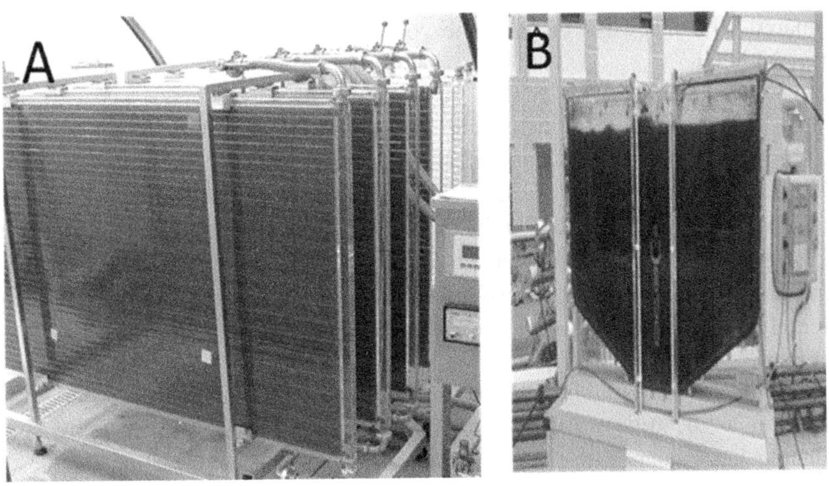

Figure 3.7. *Flat photoreactors. (A) plastic. Volume: 500 L (source: IGV Biotech – Own work, CC BY-SA 3.0, https://commons.wikimedia.org/w/index.php?curid=25767678). (B) Glass. Volume: 200 L. Graesiella sp. culture, BEA Gran Canaria Laboratory, Spain. For a color version of this figure, see www.iste.co.uk/ghoul/bioreactor.zip*

Reactor	Species	Size	Productivity	Reference
			Open-air systems	
Circular basins	*Chlorella pyrenoidosa*	0.2 m^2	15.7–18.5 $g/m^2/j$	Meng et al. (2015)
	Chlorella sp	<20 m^2	8.5–21 $g/m^2/j$	Singh and Dhar (2011)
Raceways	*Arthrospira platensis*	1.5–12 Ha	6–12 $g/m^2/j$	Vonshak and Richmond (1988)
	Dunaliella sp	250 Ha	1 $g/m^2/j$	Tredici (2004)
	Haematococcus pluvialis	1,000 L	1.44–2.10 g/m^2 /j	Borowiak et al. (2021)
	Scenedesmus obliquus	530 L	0.03 g/L/j (8.26 $g/m^2/j$)	Arbib et al. (2013)
	Muriellopsis sp.	2,000 L	12.9 ± 1.7 $g/m^2/j$	Blanco et al. (2007)
	Nannochloropsis sp.	4.73 m^2	0.03–0.08 g/L/j	De Vree et al. (2015)
Greenhouse raceways	*Spirulina platensis*	2.61 m^2	7.1–24.2 $g/m^2/j$	Pessi et al. (2022)
	Spirulina almeriensis	80 m^2	24.9 ± 0.9 $g/m^2/j$	Morillas-España et al. (2022)
	Chlorella sp.	0.986 m^2	4–21 $g/m^2/j$	Hase et al. (2000); Pessi et al. (2022)
	Dunaliella salina	2.61 m^2	3.8–13.9 $g/m^2/j$	Pessi et al. (2022)
Bubble columns	*Graesiella* sp.	100 L	0.5 g/L/ j	Gonji et col (2021)
	Picocystis sp.	20 L	0.5 ± 0.05 g /L/j	Ben Ali et al. (2022)
	Amphidinium carterae	80 L	Undefined	Morales-Amador (2021)
	Coccomyxa onubensis	800 L	0.14 g/L/j	Fuentes et al. (2020)

Reactor	Species	Size	Productivity	Reference
Thin layer	*Chlorella and Scenedesmus obliquus*	900 m^2 (5 cm) and 650 m^2 (<1 cm)	14–18 and >55 g/m^2/j	Grivalský et al. (2019)
		1,500 m^2 (<5 cm)	15–35 g/L/j	Masojídek et al. (2015)
	Closed systems (PBR)			
Vertical tubular	*Nannochloropsis* sp.	1,060 L	0.31–0.71g/L/j	De Vree et al. (2015)
	Scenedesmus obliquus	330 L	21.76 ± 0.30 g/m^2/j	Arbib et al. (2013)
	Nannochloropsis gaditana	1,020 L	0.6–0.7 g/L/j	San Pedro et al. (2014)
	A. platensis	19.6–110 L	0.2–1.4 g/L/j	Acién et al. (2019)
	Phaeodactylum tricornutum	200 L	1.2 g/L/j	Acién et al. (2013)
	Chlorella pyrenoidosa	300 L	70 g /m^2/j	Lee et al. (1996)
	Scenedesmus sp.	4,000 L	1.2–1.4 g/L/j	Fernández-Sevilla et al. (2005)
Horizontal tubular	*Nannochloropsis* sp.	98–560 L	0.30–0.85 g/L/j	De Vree et al. (2015); Zittelli et al. (2013)
	Nannochloropsis gaditana	2,500 L	0.12–0.2 g/L/j	Camacho-Rodríguez et al. (2014)
Vertical flat	*Nannochloropsis* sp.	110–1,000 L	0.16–1.70 g/L/j	Zittelli et al. (2013); Cheng-Wu et al. (2001)
Horizontal flat	*Nannochloropsis* sp.	60–200 L	0.27–1.45 g/L/j	De Vree et al. (2015); Zou and Richmond (1999)

Table 3.1. *Main types of photobioreactor used for microalgae cultivation*

However, this configuration promotes the risk of photoinhibition, or even total photooxidation, especially during strong summer solar radiation. Hu and Sommerfeld (2010) have proposed a flat reactor system consisting of several units with different layouts (vertical and inclined), different light paths (2.5–30 cm) and different geometric configurations. The units can be connected in modules and installed at different levels. This variability makes it possible to arrange the units in such a way as to optimize access to light, depending on their location in the operating area.

Despite their simple geometry, the fragile structure of flat PBRs limits their use on a large scale. Several solutions have been proposed. Horizontal or vertical deflectors (Zhang et al. 2001a, 2001b), shortened deflectors at the bottom and top (Mohr and Emminger 2009) or honeycomb channels (Tredici et al. 1991) are inserted between the walls that reinforce the structure and, at the same time, improve mixing by reducing blind spots.

The installation cost of flat reactors is lower than that of tubular PBRs (Norsker et al. 2011), but scaling up is more demanding for sustainable operation. It is limited by the risks of wall clogging, which restrict light penetration, as well as by the complexity of disinfecting and cleaning the structures between culture cycles (Wang et al. 2012).

3.2. Conclusion

It is important to note that each microalgae production system has its advantages and disadvantages. Choosing the most appropriate system depends on specific production objectives, local conditions and available resources. An integrated approach and continuous optimization are needed to overcome the technical, economic and environmental challenges associated with large-scale microalgae production.

Although it is theoretically possible to compare different reactor designs for selected strains and applications, the suitability of the cultured strain with the technical requirements (shape and agitation) of the culture system used and the environmental conditions of the culture are the key points to consider in establishing the appropriate choice. Selecting the right microalgae strains for a suitable large-scale production system is a major challenge. Some strains may be fast growing and high in compounds of interest, while others may be more resistant to harsh culture conditions. In addition, optimization of culture parameters, such as lighting, temperature, pH and nutrients, is necessary to maximize productivity.

At present, almost all the world's microalgae production is carried out using open raceways. Although these systems offer great flexibility for expansion at low set-up and running costs, their productivity, consistency of product quality and the strains suitable for cultivation are limited.

As the market moves toward species and metabolite diversification, and stricter regulations are imposed on the quality of products from the microalgae industry, the use of closed systems to control cultivation conditions is becoming increasingly essential. Closed systems guarantee high productivity and the possibility of cultivating different species, regardless of climatic conditions. However, their installation and operating costs, as well as the complexity of their scale-up and maintenance, currently limit their development.

The development of microalgae production therefore seems to depend on the balance between two antagonisms: on the one hand, the cost of the system and its simplicity of operation, and, on the other hand, optimizing the productivity and quality of the product, whether it be biomass or targeted metabolites.

3.3. References

Abdur Razzak, S., Hossain, M., Lucky, R., Bassi, A., de Lasa, H. (2013). Integrated CO_2 capture, wastewater treatment and biofuel production by microalgae culturing–A review. *Renewable and Sustainable Energy Reviews*, 27, 622–653.

Acién, F.G., Sevilla, J.M., Molina, E. (2013). Photobioreactors for the production of microalgae. *Rev. Environ. Sci. Biotechnol.*, 12, 131–151.

Acién, F.G., Molina-Grima, E., Reis, A., Torzillo, G., Chini Zittelli, G., Claudia, S., Masojídek, J. (2019). Photobioreactors for the production of microalgae. In *Microalgae-Based Biofuels and Bioproducts*, Gonzalez-Fernandez, C. and Muñoz, R. (eds). Elsevier, Amsterdam.

Andersen, R.A. (2005). *Algal Culturing Techniques*. Academic Press, New York.

Apt, K.E. and Behrens, P.W. (1999). Commercial developments in microalgal biotechnology. *J. Phycol.*, 35, 215–226.

Arbib, Z., Ruiz, J., Álvarez-Díaz, P., Garrido-Pérez, C., Barragan, J., Perales, J.A. (2013). Long-term outdoor operation of a tubular airlift pilot photobioreactor and a high rate algal pond as tertiary treatment of urban wastewater. *Ecological Engineering*, 52, 143–153.

Becker, E.W. (1994). *Microalgae: Biotechnology and Microbiology. Biotechnology*. Cambridge University Press, Cambridge.

Ben Ali, R., Ben Ouada, S., Leboulanger, C., Jebali, A., Sayadi, S., Ben Ouada, H. (2022). Emerging contaminants and nutrients recovery by *Picocystis* sp. under continuous culture in contaminated secondary municipal wastewater effluent. *Algal Research,* 66, 102804.

Benemann, J.R. and Oswald, W.J. (1994). Systems and economic analysis of microalgae ponds for conversion of CO_2 to biomass. Technical Report, September 1993–December 1993 [Online]. Available at: https://www.osti.gov/biblio/137315 [Accessed 05 July 2023].

Bernard, O. (2014). La production de micro-algues sous serre photovoltaique : évolution ou révolution ? *Revue de l'Energie*, 621, 385–392.

Blanco, A.M., Moreno, J., Del Campo, J.A., Rivas, J., Guerrero, M.G. (2007). Outdoor cultivation of lutein-rich cells of *Muriellopsis* sp. in open ponds. *Appl. Microbiol. Biotechnol.*, 73(6), 1259–1266.

Borowiak, D., Lenartowicz, P., Grzebyk, M., Wiśniewski, M., Lipok, J., Kafarski, P. (2021). Novel, automated, semi-industrial modular photobioreactor system for cultivation of demanding microalgae that produce fine chemicals. The next story of *H. pluvialis* and astaxanthin. *Algal Research*, 53, 102151.

Borowitzka, M.A. (2013). High-value products from microalgae–Their development and commercialization. *Journal of Applied Phycology*, 25(3), 743–756.

Brindley, C., Garcia-Malea, M.C., Acien, F.G., Fernandez, J.M., Garcıa, J.L., Molina, E. (2004). Influence of power supply in the feasibility of *phaeodactylum tricornutum* cultures. *Biotechnol. Bioeng.*, 87, 723–733.

Camacho-Rodríguez, J., González-Céspedes, A.M., Cerón-García, M.C., Fernández-Sevilla, J.M., Acién-Fernández, F.G., Molina-Grima, E. (2014). A quantitative study of eicosapentaenoic acid (EPA) production by *Nannochloropsis gaditana* for aquaculture as a function of dilution rate, temperature and average irradiance. *Applied Microbiology and Biotechnology*, 98, 2429–2440.

Carlsson, A.S., van Beilen, J.B., Möller, R., Clayton, D., Bowles, D. (2007). *Micro- and Macro-algae. Utility for Industrial Applications: Outputs from the EPOBIO project*. CPL Press, Newbury.

Carvalho, A.P., Meireles, L.A., Malcata, F.X. (2006). Microalgal reactors: A review of enclosed system designs and performances. *Biotechnol. Prog.*, 22(6), 1490–1506.

Cheng-Wu, Z., Zmora, O., Kopel, R., Richmond, A. (2001). An industrial-size flat plate glass reactor for mass production of *Nannochloropsis* sp. (Eustigmatophyceae). *Aquaculture*, 195(1–2), 35–49.

Chisti, Y. (2013a). Constraints to commercialization of algal fuels. *Journal of Biotechnology*, 167(3), 201–214.

Chisti, Y. (2013b). Raceways-based production of algal crude oil. *Green*, 3(3–4), 195–216.

Costa, J.A.V. and de Morais, M.G. (2014). An open pond system for microalgal cultivation. In *Biofuels from Algae*, Pandey, A., Lee, D.J., Chisti, Y., Soccol, C.R. (eds). Elsevier, Amsterdam, 1–22.

De Andrade, F.P., de Farias Silva, C.E., Medeiros, J.A., Vieira, R.C., de Sá Filho, M.L.F., Santos, G.K.S. (2022). Consortium between microalgae and other microbiological groups: A promising approach to emphasise the sustainability of open cultivation systems for wastewater treatment. *Journal of Water Process Engineering*, 50, 103211.

De Vree, J.H., Bosma, R., Janssen, M. (2015). Comparison of four outdoor pilot-scale photobioreactors. *Biotechnology Biofuels.*, 8, 215.

Doucha, J. and Lívanský, K. (1995). Novel outdoor thin-layer high-densitymicroalgal culture system: Productivity and operational parameters. *Algological Studies*, 76, 129–147.

Doucha, J. and Lívanský, K. (2006). Productivity, CO_2/O_2 exchange and hydraulics in outdoor open high-densitymicroalgal (*Chlorella* sp.) photobioreactors operated in a Middle and Southern European climate. *J. Appl. Phycol.*, 18(6), 811–826.

Doucha, J. and Lívanský, K. (2009). Outdoor open thin-layer microalgal photobioreactor: Potential productivity. *J. Appl. Phycol.*, 21(1), 111–117.

Dragone, G., Fernandes, B., Vicente, A.A., Teixeira, J.A. (2010). Third generation biofuels from microalgae. In *Current Research, Technology and Education Topics in Applied Microbiology and Microbial Biotechnology*, Mendez-Vilas, A. (ed.). Formatex, Badajoz, 1355–1366.

Egbo, M., Okoani, A., Okoh, I. (2018). Photobioreactors for microalgae cultivation–An overview. *International Journal of Scientific & Engineering Research*, 9(11), 2229–5518.

Eriksen, N.T. (2008). The technology of microalgal culturing. *Biotechnol. Lett.,* 30, 1525–1536.

Fernández-Sevilla, J.M., Acien Fernandez, F.G., Perez-Parra, J., Magán Cañadas, JJ., Granado-Lorencio F., Olmedilla, B. (2008). Large-scale production of high-content lutein extracts from S. *almeriensis*. *Proceedings of the 11th International Conference on Applied Phycology*, Galway.

Fuentes, J.L., Montero, Z., Cuaresma, M., Ruiz-Domínguez, M.C., Mogedas, B., Nores, I.G., González del Valle, M., Vílchez, C. (2020). Outdoor large-scale cultivation of the acidophilic microalga *Coccomyxa onubensis* in a vertical close photobioreactor for lutein production. *Processes*, 8, 324.

Gongi, W., Cordeiro, N., Pinchetti, J.L.G., Ben Ouada, H. (2021). Production of exopolymer substances from the thermophilic chlorophyte Graesiella: Industrial and ecological applications. *Journal of Applied Phycology*, 33, 343–356.

Grivalský, T., Ranglová, K., da Câmara Manoel, J.A. (2019). Development of thin-layer cascades for microalgae cultivation: Milestones. *Folia Microbiology*, 64, 603–614.

Guieysse, B., Béchet, Q., Shilton, A. (2012). Variability and uncertainty in water demand and water footprint assessments of fresh algae cultivation based on case studies from five climatic regions. *Bioresource Technology,* 128C, 317–323.

Guo, Z., Phooi, W.B.A., Lim, Z.J., Tong, Y.W. (2015). Control of CO_2 input conditions during outdoor culture of *Chlorella vulgaris* in bubble column photobioreactors. *Bioresource Technology*, 186, 238–245.

Harris, L., Tozzi, S., Wiley, P., Young, C., Richardson Tra-My, J., Clark, K., Trent, J.D. (2013). Potential impact of biofouling on the photobioreactors of the Offshore Membrane Enclosures for Growing Algae (OMEGA) system. *Bioresource Technology*, 144, 420–428.

Harun, R., Singh, M., Forde, G.M., Danquah, M.K. (2010). Bioprocess engineering of microalgae to produce a variety of consumer products. *Renewable and Sustainable Energy Reviews*, 14(3), 1037–1047.

Hase, R., Oikawa, H., Sasao, C., Morita, M., Watanabe, Y. (2000). Photosynthetic production of microalgal biomass in a raceway system under greenhouse conditions in Sendai city. *Journal of Bioscience and Bioengineering*, 89(2), 157–163.

Hu, Q. and Sommerfeld, M. (2010). Photobioreactor and uses therefore. U.S. Patent Application US2010/0028976 A1.

Hu, Q., Kurano, N., Kawachi, M., Iwasaki, I., Miyachi, S. (1998). Ultrahigh cell density culture of a marine green alga, *Chlorococcum littorale* in a flat plate photobioreactor. *Appl. Microbiol. Biotechnol.*, 49, 655–662.

Huntley, M.E. and Redalje, D.G. (2007). CO_2 mitigation and renewable oil from photosynthetic microbes: A new appraisal. *Mitigation and Adaptation Strategies for Global Change*, 12, 573–608.

Janssen, M., Tramper, J., Mur, L.R., Wijffels, R.H. (2003). Enclosed outdoor photobioreactors: Light regime, photosynthetic efficiency, scale-up, and future prospects. *Biotechnol. Bioeng.*, 81, 193–210.

Jerez, C.G., Navarro, E., Malpartida, I., Rico, R., Masojıdek, J., Abdala Diaz, R., Lopez Figueroa, F.L. (2014). Hydrodynamics and photosynthesis performance of *chlorella fusca* (Chlorophyta) grown in a thin-layer cascade (TLC) system. *Aquatic Biology*, 22, 1–21.

Kochem, L.H., Da Fré, N.C., Redaelli, C., Rech, R., Marcílio, N.R. (2014). Characterization of a novel flat-panel airlift photobioreactor with an internal heat exchanger. *Chem. Eng. Technol.*, 37, 59–64.

Kunjapur, A.M. and Bruce Eldridge, R. (2010). Photobioreactor design for commercial biofuel production from microalgae. *Ind. Eng. Chem. Res.*, 49(8), 3516–3526.

Lee, Y.K. (1997). Commercial production of microalgae in the Asia-Pacific rim. *J. Appl. Phycol.*, 9, 403–411.

Lee, Y.K. (2001). Microalgal mass culture systems and methods: Their limitation and potential., *J. Appl. Phycol.*, 13, 307–315.

Lee, Y.K., Ding, S.Y., Hoe, C.H., Low, C.-S. (1996). Mixotrophic growth of *Chlorella sorokinianain* outdoor enclosed photobioreactor. *J. Appl. Phycol.*, 8, 163–169.

Li, J., Stamato, M., Velliou, E. (2015). Design and characterization of a scalable airlift flat panel photobioreactor for microalgae cultivation. *J. Appl. Phycol.*, 27, 75–86.

Liao, Q., Zhong, N., Zhu, X., Huang, Y., Chen, R. (2015). Enhancement of hydrogen production by optimization of biofilm growth in a photobioreactor. *Int. J. Hydrog. Energy.*, 40, 4741–4751.

Lindblad, P., Fuente-Herraiz, D., Borbe, F., Cicchi, B., Conejero, J.A., Couto, N., Celesnik, H. (2019). CyanoFactory, a European consortium to develop technologies needed to advance cyanobacteria as chassis for production of chemicals and fuels. *Algal Research*, 41, 1–15.

Lundquist, T., Woertz, I., Quinn, N., Benemann, J. (2010). A realistic technology and engineering assessment of algae biofuel production. Report, Energy Biosciences Institute, University of California, Berkeley.

Masojídek, J., Sergejevova, M., Malapascua, J.R., Kopecky, J. (2015). Thin-layer systems for mass cultivation of microalgae: Flat panels and sloping cascades. In *Algal Biorefineries*. Prokop, A., Bajpai, R., Zappi, M. (eds). Springer, Berlin/Heidelberg, 237–261.

Mendoza, J.L., Granados, M.R., de Godos, I., Acien, F.G., Molina, E., Banks, C., Heaven, S. (2013). Fluid-dynamic characterization of real-scale raceway reactors for microalgae production. *Biomass Bioenergy*, 54, 267–275.

Meng, C., Huang, J., Ye, C., Cheng, W., Chen, J., Li, Y. (2015). Comparing the performances of circular ponds with different impellers by CFD simulation and microalgae culture experiments. *Bioprocess Biosyst. Eng.*, 38(7), 1347–1363.

Mohr, M. and Emminger, F. (2009). Method and device for photochemical process. PCT Patent, WO2009/094680.

Morales-Amador, A., Molina-Miras, A., López-Rosales, L., Sánchez-Mirón, A., García-Camacho, F., Souto, M.L., Fernández, J.J. (2021). Isolation and structural elucidation of new amphidinol analogues from *Amphidinium carterae* cultivated in a pilot-scale photobioreactor. *Mar. Drugs.*, 19, 432.

Morales-Amaral, M.M., Gomez-Serrano, C., Acien, F.G., Fernandez-Sevilla, J.M., Molina-Grima, E. (2015). Outdoor production of *Scenedesmus sp.* in thin-layer and raceway reactors using centrate from anaerobic digestion as the sole nutrient source. *Algal Research*, 12, 99–108.

Morillas-España, A., Villaró, S., Ciardi, M., Acién, G., Lafarga, T. (2022). Production of *Scenedesmus almeriensis* using pilot-scale raceway reactors located inside a greenhouse. *Phycology*, 2(1), 76–85.

Norsker, N.H., Barbosa, M.J., Vermuë, M.H., Wijffels, R.H. (2011). Microalgal production–A close look at the economics. *Biotechnology Advances*, 29(1), 24–27.

Pessi, B.A., Pruvost, E., Talec, A., Sciandra, A., Bernard, O. (2022). Does temperature shift justify microalgae production under greenhouse? *Algal Research*, 61, 102579.

Pulz, O. (2001). Photobioreactors: Production systems for phototrophic microorganisms. *Applied Microbiology and Biotechnology*, 57, 287–293.

Pulz, O. and Scheibenbogen, K. (1998). Photobioreactors: Design and performance with respect to light energy input. In *Bioprocess and Algae Reactor Technology, Apoptosis*, Scheper, T. (ed.). Springer, Berlin.

San Pedro, A., González-López, C.V., Acien, F.G., Molina-Grima, E. (2014). Outdoor pilot-scale production of *Nannochloropsis gaditana:* Influence of culture parameters and lipid production rates in tubular photobioreactors. *Bioresource Technology*, 169, 667–676.

Setlík, I., Sust, V., Malek, I. (1970). Dual-purpose open circulation units for large-scale culture of algae in temperate zones. Basic design considerations and scheme of a pilot plant. *Algological Studies/Archiv für Hydrobiologie*, 1, 111–164.

Sforza, E., Enzo, M., Bertucco, A. (2014). Design of microalgal biomass production in a continuous photobioreactor: An integrated experimental and modeling approach. *Chemical Engineering Research and Design*, 92(6), 1153–1162.

Sierra, E., Acien, F.G., Fernández, J.M., García, J.L., González, C., Molina, E. (2008). Characterization of a flat plate photobioreactor for the production of microalgae. *Chem. Eng. J.*, 138, 136–147.

Singh, N.K. and Dhar, D.W. (2011). Microalgae as second-generation biofuel: A review. *Agronomy for Sustainable Development*, 31(4), 605–629.

Slegers, P.M., van Beveren, P.J.M., Wijffels, R.H., van Straten, G., van Boxtel, A.J.B. (2013). Scenario analysis of large-scale algae production in tubular photobioreactors. *Appl. Energy*, 105, 395–406.

Sompech, K., Chisti, Y., Srinophakun, T. (2012). Design of raceway ponds for producing microalgae. *Biofuels*, 3(4), 387–397.

Stroiazzo-Mougin, B.A.J. (2011). Continuous system for converting CO_2 into high value-added and/or nutritional value products and energy resources. U.S. Patent Application US2011/0195493 A1.

Tang, H., Chen, M., Simon Ng, K.Y., Salley, S.O. (2012). Continuous microalgae cultivation in a photobioreactor. *Biotechnology and Bioengineering*, 109(10), 2468–2474.

Torzillo, G. and Zittelli, G.C. (2015). Tubular photobioreactors. In *Algal Biorefineries. Volume 2: Products and Refinery Design*. Prokop, A., Bajpai, R.K., Zappi, M.E. (eds). Springer, Cham, 187–212.

Torzillo, G., Giannelli, L., Martınez-Roldan, A.J., Verdone, N., De Filippis, P., Scarsella, M., Bravi, M. (2010). Microalgae culturing in thin-layer photobioreactors. *Italian Assoc. Chem. Eng. AIDIC.*, 20, 265–270.

Tredici, M.R. (2004). Mass production of microalgae: Photobioreactors. In *Handbook of Microalgal Culture: Biotechnology and Applied Phycology*, Richmond, A. (ed.). Wiley-Blackwell, Hoboken.

Tredici, M.R., Carlozzi, P., Zittelli, G.C., Materassi, R. (1991). A vertical alveolar panel (VAP) for outdoor mass cultivation of microalgae and cyanobacteria. *Bioresour. Technol.*, 38, 153–159.

Ugwu, C.U., Aoyagi, H., Uchiyama, H. (2008). Photobioreactors for mass cultivation of algae. *Bioresource Technology*, 99, 4021–4028.

Vasumathi, K.K., Premalatha, M., Subramanian, P. (2012). Parameters influencing the design of photobioreactor for the growth of microalgae. *Renewable and Sustainable Energy Reviews*, 16(7), 5443–5450.

Vo, H.N.P., Ngo, H.H., Guo, W., Nguyen, T.M.H., Liu, Y., Liu, Y., Nguyen, D.D., Chang, S.W. (2019). A critical review on designs and applications of microalgae-based photobioreactors for pollutants treatment. *Science of the Total Env.*, 651, 1549–1568.

Vonshak, A. and Richmond, A. (1988). Mass production of the blue-green alga Spirulina: An overview. *Biomass*, 15(4), 233–247.

Wang, B., Lan, C.Q., Horsman, M. (2012). Closed photobioreactors for production of microalgal biomasses. *Biotechnol. Adv.*, 30(4), 904–912.

Wigmosta, M., Coleman, A., Skaggs, R., Huesemann, M. (2011). National microalgae biofuel production potential and resource demand. *Water Resources Research*, 47(3), W00H04.

Zhang, K., Miyachi, S., Kurano, N. (2001a). Photosynthetic performance of a cyanobacterium in a vertical flat-plate photobioreactor for outdoor microalgal production and fixation of CO_2. *Biotechnol. Lett.*, 23, 21–26.

Zhang, K., Miyachi, S., Kurano, N. (2001b). Evaluation of a vertical flat-plate photobioreactor for outdoor biomass production and carbon dioxide bio-fixation: Effects of reactor dimensions, irradiation and cell concentration on the biomass productivity and irradiation utilization efficiency. *Appl. Microbiol. Biotechnol.*, 55, 428–433.

Zittelli, G.C., Biondi, N., Rodolfi, L., Tredici, M.R. (2013). Photobioreactors for mass production of microalgae. In *Handbook of Microalgal Culture: Applied Phycology and Biotechnology*, Richmond, A. and Hu, Q. (eds). Wiley-Blackwell, Hoboken.

Zou, N. and Richmond, A. (1999). Effect of light-path length in outdoor fiat plate reactors on output rate of cell mass and of EPA in *Nannochloropsis* sp. *Journal of Biotechnology*, 70, 351–356.

4

Bioreactor Hydrodynamics and Mixing

Céline LOUBIÈRE and Eric OLMOS
Laboratoire Réaction et Génie des Procédés (LRGP), CNRS, Université de Lorraine, Nancy, France

4.1. Introduction

Mastering the mixing operation in bioreactors is a key step in their optimization and adaptation to different types of microorganisms, such as bacteria, yeast and fungi. This operation is the subject of numerous research projects involving theoretical, experimental and modeling approaches. The main objective is to determine the configuration and operating conditions that are best suited to meet the varied needs of each type of bioreactor and each microorganism, such as homogenization, dispersion, aeration, heat transfer and substrate dissolution. The ideal agitation system must not only ensure the operations described above, but it must also be reasonably priced and preserve the integrity of the biological system used.

Different types of agitation technologies exist: mechanical, static, rotating tank, gas bubbles, etc. This chapter will focus on the technology that is most commonly encountered in bioreactors, dedicated to the production of biomass and food ingredients.

In order to properly describe the mixture in place in the bioreactor, it is necessary to identify the parameters that can affect the efficiency of this unit operation, and to

Bioreactor Implementation in the Agro-Food Industries,
coordinated by Mohamed GHOUL. © ISTE Ltd 2024.

consider the fact that bioprocesses often involve multiphase systems (gas, liquid and solid) with physicochemical properties that evolve over time.

This chapter will first define the basic concepts of agitated bioreactor hydrodynamics on a macroscopic scale, as well as the main types of flow. These will be illustrated by the typical impellers found in bioprocesses. The focus will be on multiphase mixtures, with particular reference to substrate dispersion and gas bubbles. Secondly, the nature of mixing-related phenomena will be presented at the micro-mixture scale, with particular reference to turbulent operations. Lastly, a summary of methods for characterizing the mixing efficiency of often complex biological systems will be presented.

4.2. Hydrodynamics and macroscopic mixing in bioreactors

4.2.1. *Bioreactor mixing technologies*

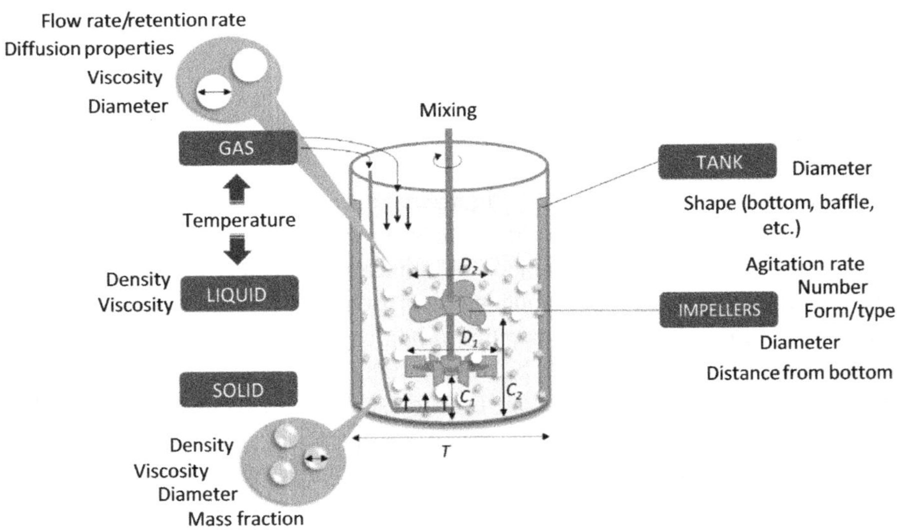

Figure 4.1. *Diagram of operating parameters affecting mixing in a mechanically stirred reactor. For a color version of this figure, see www.iste.co.uk/ghoul/bioreactor.zip*

The selection and sizing of agitators for a given bioreactor depend on the specific needs of the process to be implemented (mixing, gas transfer, suspension, etc.), the nature of the fluids involved (number of phases and types, viscosity, etc.) and potential specific constraints (fragile and shear-sensitive elements, foam production, etc.). The impact of impeller design, tank design and the typical proportions generally applied in bioreactors will be presented. The main parameters involved in mixing operations are summarized in Figure 4.1.

4.2.1.1. *Flow and types of agitation impellers*

There are three types of flow (Figure 4.2): radial, axial and tangential. The following descriptions present the most commonly used impellers in fermentative processes and reactors in general, but are by no means exhaustive (Roustan et al. 1997).

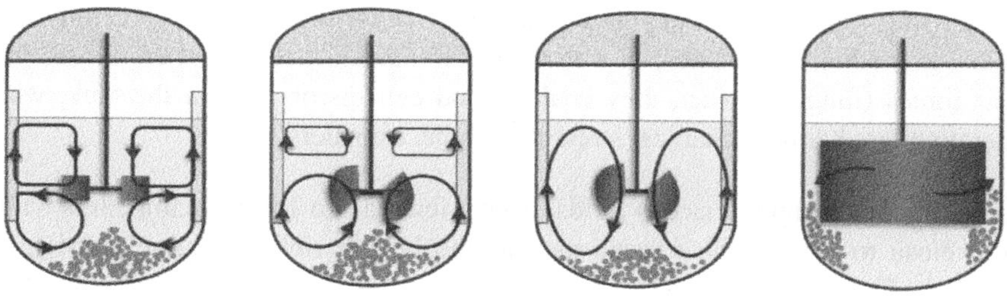

a) b) c) d)

Figure 4.2. *Diagram of circulation loops according to the type of agitator: (a) radial flow, (b) axial high-pumping flow, (c) axial low-pumping flow and (d) tangential flow. For a color version of this figure, see www.iste.co.uk/ghoul/bioreactor.zip*

Radial flows are generated by vertical, straight-bladed impellers. The most well known of these are the Rushton turbine (Figure 4.3(a)) and the straight-bladed turbine (Figure 4.3(b)). The flow field follows a rectilinear path, in line with the blades to the edge of the tank, where two circulation loops are formed, one ascending and the other descending. These impellers are generally installed in bioreactors with high oxygen and heat transfer requirements. Given their configuration, they enable gas bubbles to be finely dispersed, increasing the specific exchange surface and oxygen transfer coefficient (k_La).

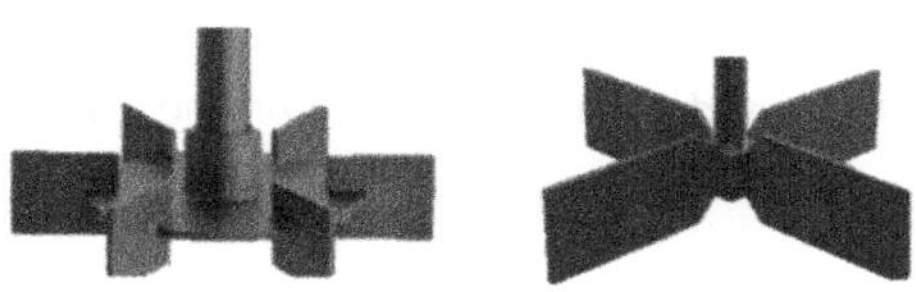

a) b)

Figure 4.3. *Examples of radial-flow impellers: (a) Rushton turbine and (b) straight-bladed impeller*

Axial flows are generated by impellers with inclined blades. Among the most common are the marine propeller (Figure 4.4(a)), the inclined-blade turbine (Figure 4.4(b)) and the elephant ear (Figure 4.4(c)). Two types of operation are possible. High pumping involves drawing fluid from beneath the impeller toward the impeller. A main circulation loop is then formed, drawing the fluid upwards at the center and pushing the fluid downwards near the walls. In some cases, a second, smaller loop may appear below, in the opposite direction. The other operation is low pumping, which corresponds to an impeller rotation in the opposite direction to high pumping, or equivalent to a mirror-symmetrical impeller design. In this case, the fluid passing through the impeller is pushed to the bottom and slides down the walls before rising again to form the circulation loop. These impellers are generally used for low to medium viscosities, as well as for biological systems that are sensitive to shear forces (fungi). In fact, they enable good cell distribution in the tank, while generating less hydromechanical stress than radial impellers.

In the case of highly viscous fluids, from a few Pa·s to 150 Pa·s, impellers with a width close to that of the vessel, such as the helical ribbon (Figure 4.4(d)), are preferred. The latter enables homogenization of the medium in a laminar flow regime.

NOTE.– Depending on the design, a greater or lesser radial component is applied in addition to the axial component. By way of comparison, the inclined-blade turbine will have a greater radial effect than the marine propeller, making it more versatile and suitable for biological systems with low shear sensitivity (yeast and bacteria).

a) b) c) d)

Figure 4.4. *Examples of axial-flow impellers: (a) marine propeller, (b) inclined-blade turbine, (c) elephant ear and (d) helical ribbon*

Tangential flows are generated by anchor-type (Figure 4.5(a)) or paddle-type (Figure 4.5(b)) impellers. These impellers encourage fluid movement as close as possible to the walls and bottom of the tank, thus promoting heat transfer and avoiding temperature gradients. Their applications generally concern fermentations carried out in viscous media or producing polysaccharides. These devices limit deposits on the walls. The anchor is used for biological systems that can tolerate operation at high shear rates, while the paddle configuration is preferred for shear-sensitive systems such as fungi.

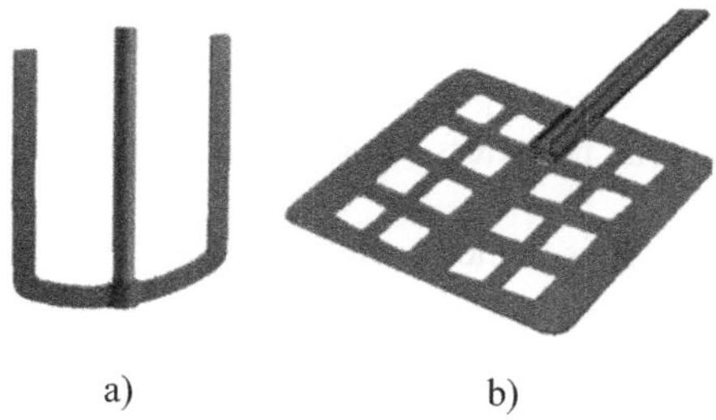

Figure 4.5. *Examples of tangential-flow impellers: (a) anchor and (b) paddle impeller*

4.2.1.2. *Fermentation tank design and standard agitator proportions*

The flow generated by the rotating impeller also depends on the obstacles encountered in the fermenter tank. The effects of the diameter, position and number of agitators, the tank geometry and the presence or absence of counter-blades (or baffles) will be discussed in detail here. The geometric parameters used to design the agitation system are shown in Figure 4.6.

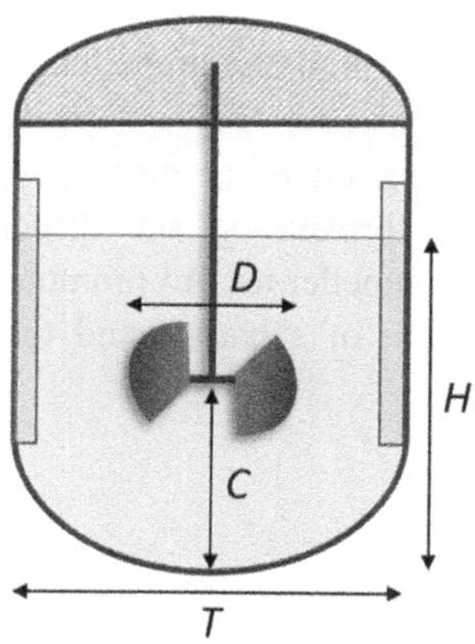

Figure 4.6. *Characteristic geometric parameters for agitation sizing. For a color version of this figure, see www.iste.co.uk/ghoul/bioreactor.zip*

The *agitator diameter D* is generally sized according to the diameter T of the bioreactor based on the D/T ratio. This ratio generally varies between 0.2 and 0.95, depending on the type of agitator and fermentation to be used. The smallest values relate to additional agitators used, so as to avoid the formation of a dead zone at the bottom of the tank beneath the main agitator. The highest values concern wall-scraping applications such as the anchor or helical ribbon. Most mixing operations applied in fermenters operate with a D/T ratio of between 1/3 and 1/2. Some applications with axial impellers can go up to a ratio of 0.7.

The *impeller's position* is, among other things, defined by C, the distance between the bottom of the tank and the bottom of the impeller's hub. Like the diameter, it is generally sized as a function of T, according to the C/T ratio. The most common situation is to operate at a value of 1/3. Beyond 1/2, the proximity of the impeller to the surface increases the risk of incorporating bubbles of overhead gas, such as the carbon dioxide released during fermentation.

The *number of agitation impellers* depends on the degree of mixing required for the microorganisms used and on the total volume. While the size of the impellers depends on the diameter of the tank T, the number of impellers depends on the height of the liquid H. It is considered necessary to add a second impeller from a height of $H = 1.2T$. This is referred to as a multi-stage configuration. Another method is based on the comparison of equation [4.1], where n is the number of impellers (You et al. 2014). The aim is to have a compromise between the mixing efficiency of the fermentation medium and the power required, but the mixing objective can also lead to using multiple impellers. A common example is the combination of a radial impeller such as the Rushton turbine with an axial impeller. In the case of mixing involving microorganisms requiring intense aeration, the turbine enables bubbles to be broken up and dispersed more effectively, thus increasing k_La, while the axial impeller acts to promote homogenization throughout the tank, avoiding the formation of substrate and temperature gradients in large fermenters.

$$\frac{H - D}{D} > n > \frac{H - 2D}{2D} \qquad [4.1]$$

Once the fluid has been set in motion by the impeller's rotation, circulation loops are formed by the proximity of the walls. The *geometry of the tank* thus has an impact on the flow. In the case of cylindrical tanks, and depending on the fermentation process used, different bottoms can be found: flat, profiled, hemispherical, etc. Profiled bottoms or hemispherical bottoms can be used in a wide range of applications. Profiled or hemispherical bottoms ensure homogeneous suspension of microorganisms and avoid cell accumulation at the bottom of the tank.

The presence of *counter-blades* or a sufficient number of plunging instruments (probes, sampling tubes, etc.) prevents the formation of a vortex, which is unfavorable to mixing and anaerobic fermentation, and promotes turbulence. Typically, two to four baffles, one-tenth the diameter of the tank *T*, are installed.

NOTE.– Other solutions avoid the need for a vortex. The agitation impeller can be offset in the fermenter, breaking the symmetry of its position. You can also adapt the design of the tank by installing a crosspiece at the bottom of the tank when the impeller is low enough, or by working in a cube-shaped tank instead of a cylinder.

4.2.2. *Macroscopic flow fields*

The energy balance in the bioreactor establishes that the required power P_{req} must be equal to the dissipated power *P*. This amount of power corresponds to the power delivered by the agitator, after deduction of losses due to friction and transmission between the motor and agitator. It is then converted into turbulent inertia and ultimately dissipated in the form of dissipated heat at the small vortex scales.

4.2.2.1. *Transport phenomena*

The mechanical energy generated by the rotation of the impeller is converted into turbulent kinetic energy, which is ultimately dissipated as heat at the micro-scale. Three main phenomena are involved at different scales.

At the bioreactor scale, fluid transport between the different zones of the bioreactor is referred to as *distribution* because of the currents induced by the agitation impeller. This phenomenon becomes slower as the volume or viscosity of the fermenting medium increases. This is reflected in higher mixing times at larger scales. Nonetheless, the agitation must be designed to be sufficient to maintain movement and renewal throughout the entire tank volume, within a reasonable time.

At the local scale, *turbulent dispersion* corresponds to the transport and distribution of fluid within the volume plots described above because of the vortices created by agitation. An increase in agitation generates smaller vortices that promote local mixing.

Lastly, at the molecular scale, *diffusion* is defined as the migration of molecules by concentration gradient or Brownian motion. Diffusion is generally slower than dispersion in agitated bioreactor mixtures, and it is therefore often neglected. It can, however, be considered rapid over short distances, and becomes significant as we move away from the ideal case of the perfectly agitated bioreactor (e.g. the appearance of poorly mixed zones on large scales, or when viscosity is high).

4.2.2.2. *Flow regime*

The flow regime depends on the properties of the medium to be mixed (density and viscosity), and the ratio between the viscous and kinetic forces at play when the medium is agitated.

Under *laminar flow* conditions, and therefore in the absence of turbulent dispersion, material and heat transfer are essentially guided by diffusion/conduction mechanisms. These transport phenomena become dominant at high viscosities. In contrast, in the case of fully established *turbulence*, flow mainly depends on inertial forces.

Turbulence can be defined as "chaotic" (aperiodic), unsteady motion in which transported quantities (mass, momentum and concentration) fluctuate in time and space. It is characterized by a set of identifiable structures: turbulent vortices, whose characteristic size varies in space. These structures are unstable and exchange matter with one another, thus promoting transfers (matter, heat and mixing). Due to friction between the fluid medium and the walls, the transition between laminar and turbulent flows occurs at a distance from the boundary layer. At the bioreactor scale, the adimensional Reynolds number is introduced to determine the flow regime. In the case of stirred fermenters, it is calculated according to equation [4.2], with ρ_L being the density (kg·m^{-3}), μ_L being the dynamic viscosity (Pa·s) of the fluid, D being the diameter of the agitation impeller (m) and N being the stirring frequency (rps).

$$Re = \frac{\text{Inertia forces}}{\text{Viscous forces}} = \frac{\rho_L N D^2}{\mu_L} \qquad [4.2]$$

The flow regime can then be estimated according to the order of magnitude of this number (Figure 4.7):

– $Re < 10$: laminar regime;

– $10 < Re < 10^3$–10^4: transient regime;

– $Re > 10^3$–10^4: turbulent regime.

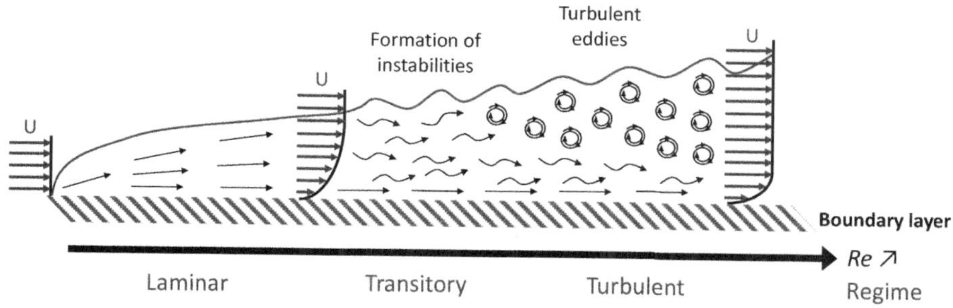

Figure 4.7. *Flow regime and turbulence evolution according to the Reynolds number. For a color version of this figure, see www.iste.co.uk/ghoul/bioreactor.zip*

4.2.3. *Quantifying mixing capacities*

4.2.3.1. *Power dissipation and power number*

Knowledge of the power dissipated is essential for optimal sizing of the mixing system. A compromise must be found to achieve sufficient power dissipation to meet mixing objectives, while minimizing investment and operating costs.

To enable comparison between two configurations, and irrespective of working scales, it is customary to work from the power density P/V, where the power is reduced to the volume of the fluid conveyed. Also known as specific power, values generally range from 10 $W \cdot m^{-3}$, for the most fragile or highly viscous systems, to up to 3 $kW \cdot m^{-3}$ for systems with a high oxygen demand.

Power is generally related to the power number Np, defined by equation [4.3], linking friction forces to inertia forces.

$$Np = \frac{P}{\rho_L N^3 D^5} \quad [4.3]$$

In the laminar regime and for Newtonian-type fermentation media, this number is inversely proportional to the Reynolds number according to the agitation constant K_p (equation [4.4]), while in the turbulent regime, N_p becomes constant and serves as a reference for comparing agitation impellers with one another.

$$Np = \frac{K_p}{Re} = \frac{K_p \mu_L}{\rho N D^2} \quad [4.4]$$

N_p depends on the impeller's geometric characteristics. In the turbulent regime, axial impellers have power numbers generally between 0.2 and 2, and are therefore lower than those of radial impellers, between 3 and 6.5. For both configurations, an increase in the D/T ratio or a decrease in the C/T ratio leads to an increase in N_p (Chapple et al. 2002).

4.2.3.2. *Pumping rate and pumping number*

The pumping flow rate Q_p represents the fluid flow rate through the volume occupied by the agitation impeller. It is proportional to the product of the characteristic velocity $N \times D$ and the characteristic surface D^2, in other words, $N \times D^3$. On this basis, we define the pumping number N_{Qp} as the ratio presented in equation [4.5], which is constant for each agitation impeller in the turbulent regime.

$$N_{Qp} = \frac{Q_p}{ND^3} \quad [4.5]$$

NOTE.– The contrast of pumping rate to power number can be a criterion for selecting an impeller according to mixing objectives, through the N_{QP}/N_p ratio. Indeed, some impellers require more power for a given pumping rate, such as radial impellers (e.g. Rushton).

4.2.3.3. *Mixing and circulation times*

In the case of stirred reactors aiming for relatively rapid homogeneity, the mixing time t_m can be defined as the time required to reach a given degree of homogeneity. For example, for a degree of homogeneity of 95% following the addition of a tracer as close as possible to the stirrer, $t_m = t_{95}$ will be reached when the time for temporal fluctuations in concentration is within plus or minus 5% of the mean concentration value (Figure 4.8).

The mixing time is related to the circulation time t_c by the circulation number N_θ according to equation [4.6]. t_c corresponds to the time of each peak period before homogenization, in other words, the time it takes for the fluid to return to the same point, while N_θ indicates the number of circulations required to achieve the targeted degree of homogenization.

$$N_\theta = \frac{t_m}{t_c} \quad [4.6]$$

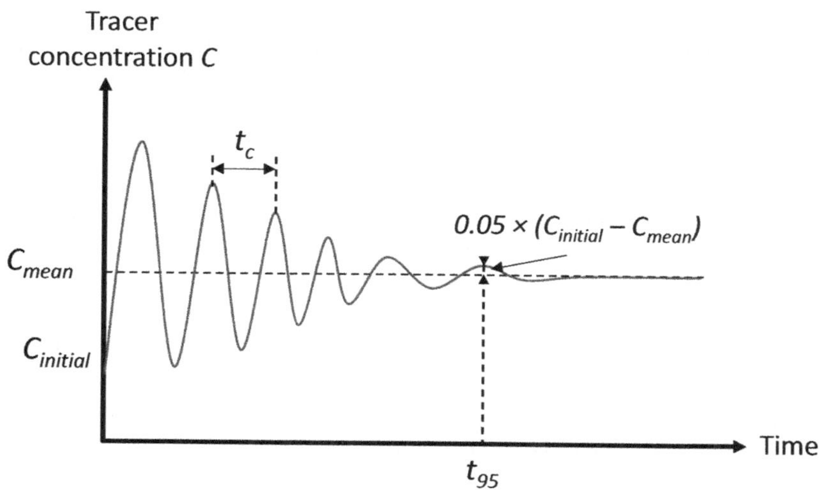

Figure 4.8. *Changes in local tracer concentration after a single addition. For a color version of this figure, see www.iste.co.uk/ghoul/bioreactor.zip*

Note that the mixing time can also be related to the Reynolds and power numbers by the mixing number $N \times t_m$. For a degree of homogenization of 95%, we get equations [4.7] and [4.8].

In the transient regime, in the intermediate sense between laminar and turbulent regimes:

$$N \times t_{95} = 33{,}500\left(\frac{D}{T}\right)^{-2} Re^{-1} N_p^{-2/3} \quad [4.7]$$

In the turbulent regime:

$$N \times t_{95} = 5.2 T^{1.5} H^{0.5} D^{-2} N_p^{-\frac{1}{3}} \quad [4.8]$$

4.2.4. *Liquid–solid mixing*

In food bioprocesses, the liquid–solid mixing operation is encountered during the substrate dissolution phase, as well as in fermentative processes involving immobilized microorganisms, or in the form of pellets or spores (fungi). In these cases, fermentative medium-particle mixing processes are used to enable dispersion, suspension, dissolution or adsorption (Paul et al. 2004; Kresta et al. 2015). As a general rule, the solid phase has a higher density than the liquid phase. It is therefore necessary to provide sufficient power to allow the solid to be suspended in order to maximize liquid–solid contact, or even to homogenize the mixture. The force involved must be greater and compensate for the sedimentation force of the particles. If, on the other hand, the solid has a lower density than the liquid phase, agitation can be used to introduce particles into the continuous phase (Mersmann et al. 1998).

To meet these objectives, we are looking for impellers that deliver good pumping capacity while ensuring a medium level of shear. This makes pitched-blade turbines (PBT) excellent candidates for these applications. For bioprocesses requiring high shear, radial impellers such as the Rushton turbine are preferred.

4.2.4.1. *Stokes number*

The Stokes number compares the kinetic energy of a particle with the energy resulting from viscous dissipation. It is used to characterize the behavior of a particle transported within the fluid. It is calculated according to equation [4.9], with $\rho_p$being the density (kg·m^{-3}) and d_p being the diameter (m) of the solid, v being the fluid velocity (m·s^{-1}) and μ being the dynamic viscosity of the fluid (Pa·s).

$$St = \frac{1}{18}\frac{\rho_S d_p v_L}{\mu_L} \qquad [4.9]$$

A Stokes number *St* << 1 implies a viscous regime. The particles follow the fluid's flow lines and are therefore transported at the same speed as the liquid. This is known as one-way coupling.

A Stokes number *St* >> 1 implies an inertial regime. In this case, particle motion mainly depends on particle inertia, and the trajectory is little influenced by the fluid. Particle displacement can then also have an impact on fluid flow. This is known as two-way coupling. Particle–particle or particle–wall collisions can also be observed. When particles collide with one another, this is then known as four-way coupling.

4.2.4.2. *State of just-suspended microorganisms*

Complete suspension of particles such as microorganisms or fungal spores is generally sought to maximize particle-medium contact, while minimizing the power required. The agitation rate required to achieve this state of complete suspension is generally referred to as N_{js} (Figure 4.9). Numerous methods have been exhaustively developed by Tamburini (2012) to estimate N_{js}. They fall into four categories:

– *Direct methods* such as Zwietering's visual method (Zwietering 1958; Nienow 1968; Arvinth et al. 1996), where this state is considered to be reached when particles remain on the tank bottom for no more than one or two seconds.

– *Indirect methods* such as the cloud height method (Kraume 1992), concentration determination by light attenuation (Delafosse et al. 2018) or the power number method (Raghava et al. 1988).

– *Numerical methods* based on data recovered from computational fluid dynamics (CFD) simulations (Kee et al. 2002; Tamburini et al. 2011).

– *Theoretical methods* based on the assumption of equilibrium between sedimentation and particle uplift (Baldi 1978; Wichterle 1988), leading to models derived from experimental results, are limited by particle properties and the geometry of the tank and agitator. The most commonly used equation is the Zwietering relation, equation [4.10] (Zwietering 1958). S is a dimensionless parameter related to the geometry of the tank and the impeller independent of the scale of the system, υ is the kinematic viscosity of the fluid, g is the gravity constant, ρ_p is the particle density, ρ_L is the fluid density (kg·m^{-3}), X is the solid mass faction, d_p is the particle diameter (m) and D is the impeller diameter (m).

$$N_{js} = S \cdot \upsilon^{0.1} \cdot \left(\frac{g(\rho_p - \rho_L}{\rho_L}\right)^{0.45} \cdot X^{0.13} \cdot d_p^{0.2} \cdot D^{-0.85} \qquad [4.10]$$

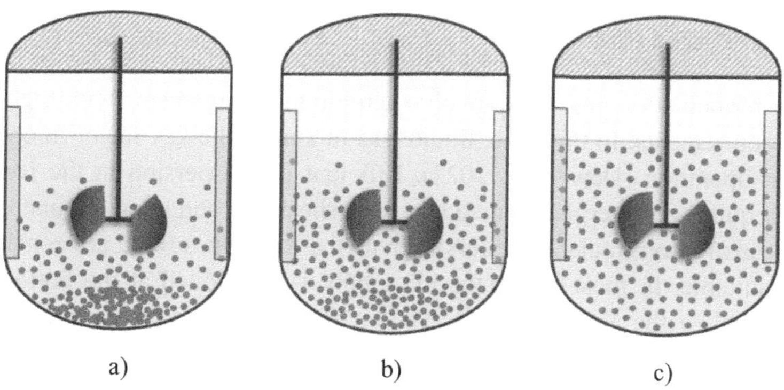

Figure 4.9. *Degree of suspension of a liquid–solid mixture: (a) partial suspension, (b) just suspension and (c) uniform suspension. For a color version of this figure, see www.iste.co.uk/ghoul/bioreactor.zip*

The agitation rate N_{js} therefore depends on the geometric dimensions of the stirrer. To reduce the value of this threshold frequency, we can increase the diameter of the impeller D or bring the impeller closer to the bottom of the tank (reduce C). Fluid–solid interactions also have an impact on this parameter. An increase in the concentration of cells in bioreactors (particles) or in the Archimedes number tends to increase the value of N_{js}.

4.2.4.3. *Degree of homogeneity in bioreactors*

Some fermentation processes involve the nonhomogeneous suspension of solids, microorganisms or spores in the tank. This is particularly the case when working with just suspension. This homogeneity can be characterized by a suspension quality factor H_s, proposed by Mak (1992), presented in equation [4.11]. This parameter represents the relative deviation between the local particle concentration and the average concentration, which would be uniform in the case of a complete and homogeneous suspension, with V_{tot} being the total volume, α_S being the local solid volume fraction and $\langle\alpha_S\rangle$ being the average volume fraction in the tank. Nevertheless, calculating this parameter requires very specific techniques to obtain local concentration values (e.g. light attenuation technique, prediction by CFD modeling, etc.).

$$H_S = 1 - \sqrt{\frac{1}{V_{tot}} \cdot \iiint_{V_{tot}} (\alpha_S - \langle\alpha_S\rangle)^2 dV} \qquad [4.11]$$

4.2.5. *Liquid–gas mixing*

Many fermentation processes involve aerobic microorganisms with high oxygen transfer requirements. In this case, liquid–gas mixing is the key factor in optimizing bioreactor operation (Dew et al. 2022). Efficient gas dispersion in the fermenting medium requires a stirrer that generates a high shear rate and good circulation in the tank. For this reason, radial impellers such as the Rushton turbine are considered to be more suitable. In most cases, the air supply is positioned at the bottom of the fermenter, below the agitation itself. The air flow is injected via a tube and exits through perforations in the tube. The number, diameter and spacing of the perforations can be used to determine the initial bubble size.

4.2.5.1. *Bubble dispersion regime and power dissipation*

Several flow regimes can be distinguished, depending on the design of the impeller and the operating conditions (agitation rate, injected air flow rate). As shown in Figure 4.10, too little agitation or an oversized air flow can lead to the clogging of the agitator. In this example, air is poorly dispersed, leading to significant dissolved oxygen gradients within the tank, and even depletion zones. In addition, the presence of gas cavities and bubbles affects flow and power dissipation. When the air flow rate is increased at a given agitation frequency, there is a gradual transition from a vortex regime to a large-cavity regime, until a single, large-scale interface is created, driven by the impeller blades. On the one hand, the gas-liquid contact area is reduced and, on the other hand, the drop in local dispersion density leads to a reduction in power dissipation, resulting in longer mixing times and lower gas–liquid transfer capacities.

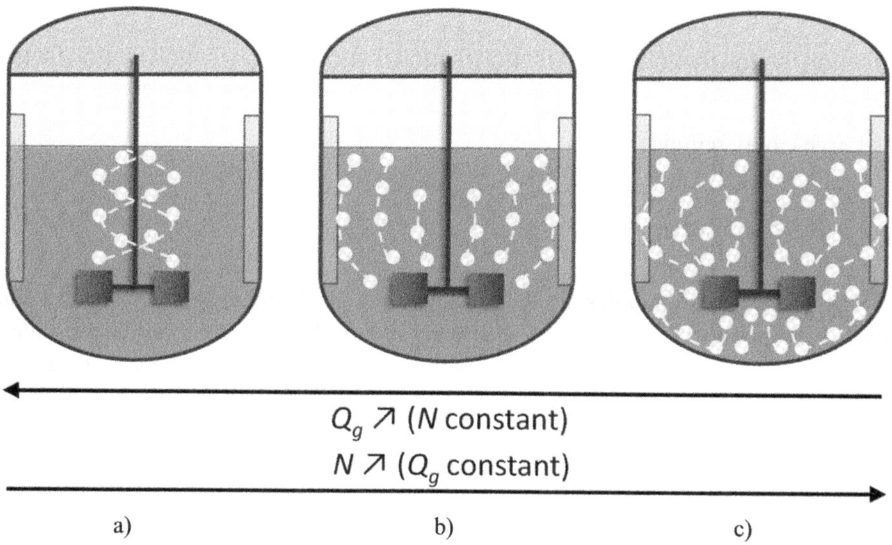

Figure 4.10. *Gas dispersion regimes as a function of gas flow Qg and stirring frequency N: (a) clogging, (b) transition and (c) complete dispersion. For a color version of this figure, see www.iste.co.uk/ghoul/bioreactor.zip*

The aeration number N_A is established according to equation [4.12], with Q_g being the gas flow rate ($m^3 \cdot s^{-1}$). From its value and the Froude number defined in equation [4.13], it is possible to predict the expected flow regime. Operating regime correlations depend on the design of the impeller. For a fermenter equipped with a Rushton turbine, equations [4.14]–[4.16] can be used with the condition of a Froude number $Fr < 0.04$. For small Froude numbers, it is assumed that the agitator has no influence on gas dispersion.

$$N_A = \frac{Q_g}{ND^3} \qquad [4.12]$$

$$Fr = \frac{N^2 D}{g} \quad [4.13]$$

Complete gas dispersion regime:

$$N_A > 0.2 \cdot \left(\frac{D}{T}\right)^{0.5} \cdot Fr^{0.5} \quad [4.14]$$

Agitator clogging regime:

$$N_A > 30 \cdot Fr \cdot \left(\frac{D}{T}\right)^{3.5} \quad [4.15]$$

Large cavity regime:

$$N_A > 0.025 \cdot \left(\frac{D}{T}\right)^{-0.5} \quad [4.16]$$

The presence of bubbles in the agitator area leads to a local change in the physical properties of the mixture. The power required to move the fluids, presented in section 4.2.3.1, decreases. P_g is the power dissipated in the presence of air and P_0 is the power in the non-aerated condition (equation [4.3]). The ratio of these two powers constitutes the relative power demand P_g/P_0. Correlations such as the Calderbank correlations allow this parameter to be estimated from the aeration number (equations [4.17] and [4.18]) (Calderbank 1958).

$$\frac{P_g}{P_0} = 1 - 12.6 \cdot N_A \; si \; N_A < 0.035 \quad [4.17]$$

$$\frac{P_g}{P_0} = 0.62 - 1.85 \cdot N_A \; si \; N_A > 0.035 \quad [4.18]$$

4.2.5.2. *Aeration rates and bubble sizes*

An important parameter for characterizing the liquid–gas mixture is the gas retention rate, representing the gas volume fraction according to equation [4.19]. It is correlated according to power laws to the dissipated power and the gas surface velocity u_g. In equation [4.20], the constants a and b depend on the properties of the liquid phase and the design of the tank and agitator.

$$\epsilon_g = \frac{V_g}{V_L + V_g} \quad [4.19]$$

$$< \varepsilon_g > \propto \left(\frac{P_g}{V}\right)^a \cdot u_g^b \qquad [4.20]$$

The air (gas) velocity is calculated as the ratio between the air flow rate and the cross-sectional area of the tank. In the case of a cylindrical tank with diameter T, this gives equation [4.21].

Note that in fermentation, gas flow can also be expressed in vvm (volumetric flow of gas per liquid volume per minute), obtained by equation [4.22] and related to surface velocity according to equation [4.23], where H is the liquid height.

$$u_g = \frac{4Q_g}{\pi T^2} \qquad [4.21]$$

$$vvm = \frac{60Q_g}{\left(\frac{\pi}{4}\right) T^2 H} \qquad [4.22]$$

$$u_g = vvm \cdot \frac{H}{60} \qquad [4.23]$$

From these data, it is possible to estimate the diameter of the bubbles. This depends on the agitator design and air supply technology selected, the physicochemical properties of the fermenting medium (viscosity, interfacial tension) and the operating conditions, including agitation rate and air flow rate. This should be considered as a global estimate. Breakdown and coalescence phenomena lead to heterogeneous distribution in the fermenter, both locally and macroscopically.

This geometric knowledge of the bubbles enables us to define the interfacial area a (m^{-1}) as the ratio of the liquid–air interface surface area to the volume of liquid in the fermenter (equation [4.24]). For a given flow rate, the smaller the bubbles, the larger a will be.

$$a = \frac{S_{interfaces}}{V_{liquid}} \qquad [4.24]$$

The diffusion velocity k_L is also defined according to equation [4.25], with D being the diffusivity ($m^2 \cdot s^{-1}$) and δ being the boundary layer thickness (m). In the context of mixing operations, the impact of agitation and viscosity will be highlighted. An increase in agitation tends to decrease δ and therefore increase k_L, while an increase in liquid viscosity tends to decrease diffusivity, and therefore k_L as well.

$$k_L = \frac{D}{\delta} \qquad [4.25]$$

Obtaining values for a and k_L is difficult, due to bubble size distribution that is subject to breakage and coalescence phenomena, and limited knowledge of oxygen diffusivity in culture media. However, it is common practice to work with the product of these two parameters, k_La, known as the volume coefficient of oxygen transfer. The latter is correlated with the previously defined power density and gas surface velocity, according to power laws presented in equation [4.26]. α and β are constants that only depend on design and technology (tank, impeller, aeration system) and A is a function of the medium. Typical values for α are between 0.3 and 0.7, and are around 0.5 for β for standard configurations. An advantage of this parameter and correlation is independence from system size, enabling its use for extrapolation of a process requiring gas transfer. Experimentally, it is relatively simple to determine k_La from methods based on oxygen concentration measurements in the fermenter.

$$k_L a = A\left(\frac{P_g}{V}\right)^{\alpha} u_g^{\beta} \qquad [4.26]$$

4.3. From macromix to micromix

The micromixing scale was introduced in the late 1950s by Danckwerts (1958) and Zwietering (1959). It is defined as the mixing process that ensures homogeneity at the microscopic and molecular scale, and corresponds to phenomena characterizing the so-called fine texture of the mixture (Villermaux 1993). The degree of micromixing lies between the following ideal boundary states:

– If the volume in the tank divides into volume plots that remain distinct from one another, without mixing, we then speak of total segregation. Molecules remain grouped together in a given plot without coming into contact and reacting with molecules from another plot as they pass through the reactor. These plots are called segregation domains.

– If, on the other hand, two molecules taken at random from the tank are always capable of coming into contact with one another and react with one another, we speak of a perfectly micromixed fluid.

Laminar mixing is governed by molecular diffusion and advection, which are regarded as slow mechanisms. In mechanically stirred reactors, this mainly concerns the mixing of highly viscous fluids. In the majority of cases, we operate in a turbulent regime, enabling us to reduce the size of the fluid's elements and thus promote mixing and the transfer of matter and heat. In the remainder of this chapter, we will concentrate on this second situation, where turbulence phenomena will be defined at the microscale with the transport of turbulent kinetic energy.

4.3.1. *Fluctuations and dissipation rates of turbulent kinetic energy*

4.3.1.1. *The turbulent cascade mechanism*

In fermentative processes, turbulence affects both the transfer of matter, particularly oxygen and CO_2, and the physiological state of the microorganisms used, especially if they are sensitive to shear forces. Characterizing turbulence in a bioreactor is therefore an important factor in optimizing its operation. This knowledge is also used to develop the methodology for extrapolation between bioreactor scales.

In 1920, Lewis Fry Richardson qualitatively explained the phenomenon of turbulence through the formation of vortices of different sizes. Agitation generates large eddies defined by a characteristic length L, a characteristic velocity U and a characteristic time $\tau_L = L/U$. The corresponding Reynolds number is expressed as $Re_L = UL/\mu$. For these large eddies, viscous forces are kept negligible. Richardson suggested that in turbulent flows, these large eddies divide into smaller eddies, with a transfer of kinetic energy. These eddies continue to divide until they become small enough for viscous forces to become significant. At this point, the kinetic energy associated with turbulence dissipates as heat.

We define the kinetic energy dissipation rate ε as a function of the large eddies according to equation [4.27].

$$< \varepsilon > = \frac{U^3}{L} \tag{4.27}$$

By conserving the total mechanical energy within an agitated tank, this dissipation rate can also be related to the power required according to equation [4.28].

$$< \varepsilon > = \frac{1}{\rho_L} \cdot \left(\frac{P}{V}\right) = \frac{N_p \cdot N^3 \cdot D^5}{V} \tag{4.28}$$

At the same time, turbulent kinetic energy k is defined as the kinetic energy of velocity fluctuations. The expression of the instantaneous fluid velocity can be described by a statistical velocity distribution, which can be characterized by the sum of two components: the time average $\bar{u}$ and the turbulent fluctuation u', according to equation [4.29]. The second component is then used to estimate the turbulent kinetic energy according to equation [4.30].

$$u_i = \bar{u}_i + u_i' \tag{4.29}$$

$$k = \frac{1}{2}\overline{u_i' u_i'} \quad [4.30]$$

The kinetic energy contained in large eddies is considered equal to the sum of the kinetic energies contained in all generated eddies, until it dissipates as heat, as shown in Figure 4.11.

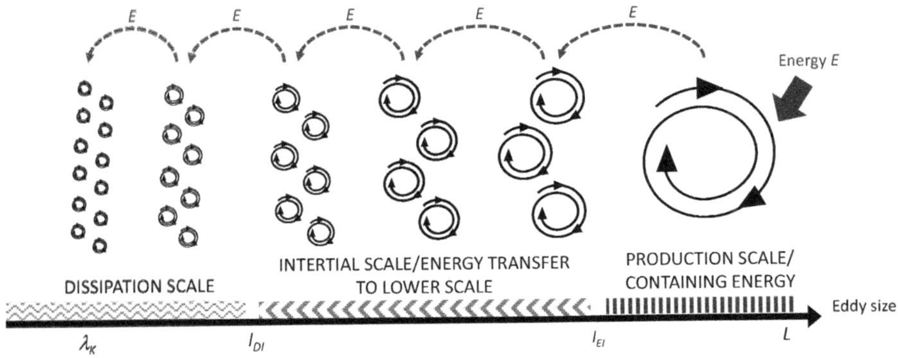

Figure 4.11. *Energy transfer from large eddies to small eddies according to turbulent cascade theory. IDI: length of transition to the dissipation scale. IEI: length of transition between large eddies (anisotropy) and small eddies (isotropy). For a color version of this figure, see www.iste.co.uk/ghoul/bioreactor.zip*

4.3.1.2. *Kolmogorov's theory*

The laws relating to the cascade mechanism were obtained by Komogorov (1941, 1991). They are based on a local isotropy hypothesis and two similarity hypotheses for sufficiently high Reynolds numbers:

– H0: small-scale turbulent motions are isotropic (Figure 4.11): $l \ll L$.

– H1: small-scale turbulent motions only depend on kinematic viscosity ν and kinetic energy dissipation rate ε (Figure 4.11): $l < l_{EI}$.

– H2: small-scale turbulent motions now only depend on ε (Figure 4.11): $\lambda_K \ll l < L$.

The first hypothesis of similarity (H1) allows us to establish the characteristic quantities of equations [4.31]–[4.33] and the associated Reynolds number equation

[4.34]. They correspond to the Kolmogorov scale of dissipation relating to the smallest eddies in the flow where energy is dissipated.

$$\text{Length: } \lambda_K = \left(\frac{\nu^3}{\varepsilon}\right)^{1/4} \quad [4.31]$$

$$\text{Rate: } u_{\lambda_K} = (\varepsilon \cdot \nu)^{1/4} \quad [4.32]$$

$$\text{Time: } \tau_{\lambda_K} = \left(\frac{\nu}{\varepsilon}\right)^{1/2} \quad [4.33]$$

$$Re_{\lambda_K} = \left(\frac{\lambda_K \cdot u_{\lambda_K}}{\nu}\right) = 1 \quad [4.34]$$

4.3.2. *Description and calculation of hydromechanical stresses*

In order to describe hydromechanical stresses, let us assume a mixture involving a suspension of microorganisms (particles) in the bioreactor. Here, we are interested in describing the stresses placed on the microorganisms by the shear generated by the liquid phase. Henzler described particle stress as the consequence of a velocity gradient in the liquid phase at the particle scale (Henzler 2000). In laminar flow, particle stress is calculated from the fluid velocity u according to Newton's equation (equation [4.35]). In turbulent flow, it is calculated from the turbulent fluctuation component of the fluid velocity u', according to equation [4.36]. Its determination depends on the turbulent energy dissipation rate ε and the Kolmogorov length λ_K, enabling the turbulence scale to be defined.

$$\tau = \mu_L \frac{du}{dx} \quad [4.35]$$

$$\tau_t = \rho_L \sqrt{u'^2} \quad [4.36]$$

4.3.3. *Spatial distribution of hydrodynamic parameters in agitated tanks*

The distribution of velocities within the fermentation tank allows us to establish the overall circulation, as well as identify potential dead zones and areas with high velocity gradients. In the case of agitated bioreactors, the lowest velocities often appear in the upper zone of the tank, close to the free surface, as this zone is further away from the impeller.

In the same way, the distribution of the turbulent kinetic energy dissipation rate ε can be established. Once again, a distinction is made between the zone around the agitator and the rest of the bioreactor. Shear is significantly higher in the agitation zone, with a turbulent energy dissipation rate up to 100 times the mean value, whereas only a quarter of this mean value can be found in the rest of the tank (Zhou et al. 1996).

From the values of ε, we can determine the size of the smallest eddies λ_K according to equation [4.31]. A first approximation is to use the power dissipated in the tank to calculate a mean turbulent kinetic energy dissipation rate $< \varepsilon >$. The associated Kolmogorov length can be used to estimate the average size of the smallest eddies in the tank. If local information is available, in particular through numerical fluid mechanics modeling, it will be possible to highlight the values of λ_K corresponding to the ε that is calculated locally in different areas of the tank. The distribution of these Kolmogorov lengths can then be constructed and the actual value of the smallest eddy recovered. These data are particularly useful in fermenters that require intense oxygen transfer, and where it is necessary to control shearing in the tank for sensitive products.

NOTE.– Optimization and extrapolation processes need to consider the fact that each molecule or element dispersed by agitation follows a path with variable local properties. The reconstruction of these paths constitutes the "lifelines". For example, for the same power dissipated in two reactors with different impellers, a particle may travel a path with variable but moderate shear intensities in one, and a predominantly low-intensity but occasionally severe path in the other. The ability to withstand these intensity thresholds and their frequency can have an impact on more sensitive mixtures.

4.4. Bioreactor mixture characterization method

Various experimental and modeling techniques for characterizing mixing in bioreactors have been described in the literature. The main ones will be briefly presented in this section, with a focus on the methods used to measure mixing time/residence time distribution (RTD) and characterize velocity fields.

Parameter	Phases involved			
	L	LS	LG	LSG
Torque ω (N·m)	×	×	×	×
Power loss P (W) or power density P/V ($W \cdot m^{-3}$)	×	×	×	×
Mixing time t_m (s) or DTS	×	×	×	×
Flow/circulation loops	×	×	×	×
Velocity fields	×	×	×	×
Local particle fraction α_s (-)		×		×
Retention rate ε_g			×	×
Suspension quality H_S		×		×
Agitation rate just for suspension N_{js} (rps)		×		×
Gas–liquid transfer coefficient k_La (s^{-1})			×	×
Gas dispersion regime			×	×

Table 4.1. *Parameters characterizing mixing and hydrodynamics in a stirred reactor (L: liquid, S: solid, G: gas)*

4.4.1. *Residence time and mixing time*

In continuous bioreactors, the RTD is a statistical method based on the time spent in the tank by a given number of inert tracer elements introduced at the inlet. A quantity of inert tracer that does not affect the flow is thus injected at the inlet and quantified at the outlet with its "age". A saline solution and conductometry are most often used, but dyes (colorimetry) or radioactive molecules (radioactivity) can also be used.

From the measured residence times, we construct a statistical frequency distribution $E(t)$ (the sum of the frequencies, in the sense of probability, is 1). This can then be used to extract several pieces of information characterizing the flow in the bioreactor.

If the average residence time $\bar{t}$ (equation [4.37]) is equal to the throughput time $\tau = V/Q$, then we are approaching the ideal case of the perfectly stirred reactor (Figure 4.12(a)).

$$\bar{t} = \int_0^\infty t \cdot E(t)dt \qquad [4.37]$$

The variance, or second central moment (equation [4.38]), provides information on the dispersion of output frequencies.

$$\sigma^2 = \int_0^{\infty} (t - \bar{t})^2 \cdot E(t)dt \qquad [4.38]$$

The third and fourth central moments specify the asymmetry of the function and the flattening of the $E(t)$ function, respectively.

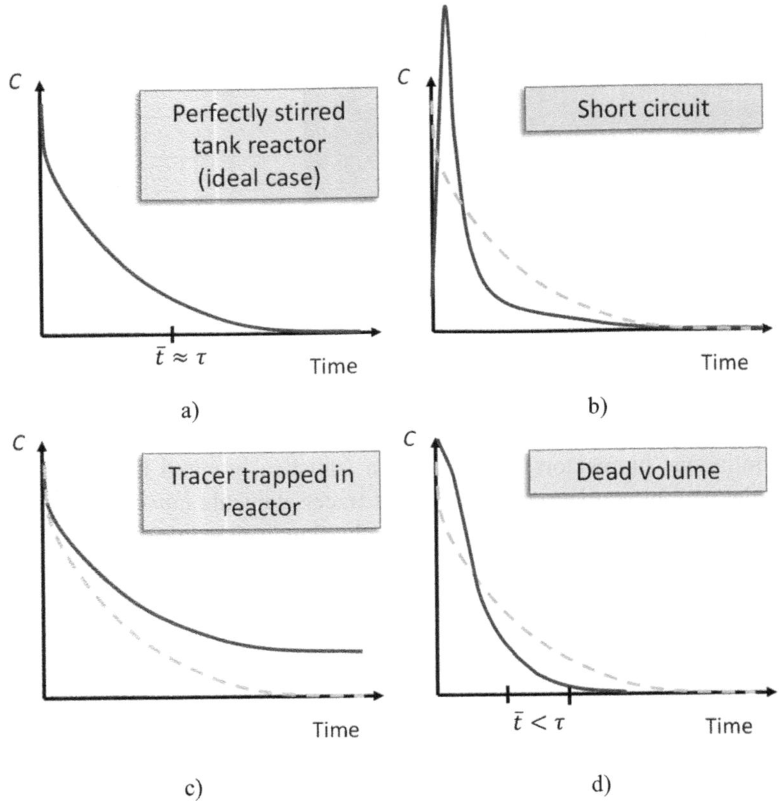

Figure 4.12. *Typical RTD profiles in an open reactor. For a color version of this figure, see www.iste.co.uk/ghoul/bioreactor.zip*

By plotting frequencies against time, we can identify deviations from ideality, such as the presence of short circuits (Figure 4.12(b)), areas where the tracer remains trapped (Figure 4.12(c)) or dead volumes (Figure 4.12(d)). In the case of a perfectly stirred reactor, the RTD expression is:

$$E(t) = \frac{1}{\tau}\exp\left(-\frac{t}{\tau}\right) \quad [4.39]$$

In closed or batch bioreactors, mixing can be characterized by the mixing time t_m, described in section 4.2.3.3. This parameter is obtained experimentally by injecting a tracer and measuring its concentration at one or more points in a different zone. The mixing time is considered to have been reached when the concentration in the measured zone reaches the homogeneity target, such as 95% of the mean concentration for t_{95}.

Acid or base (pH measurement), salt solution (conductometry) and dye (colorimetry) are most often used. Measurements are made using a probe measuring the change of concentration in a single point, or more advanced techniques such as electrical resistance tomography (ERT) enabling measurements on volume, or planar laser-induced fluorescence (pLIF) enabling measurements on planes (Ascanio 2015). pLIF involves photochromic dyes, making particles fluoresce when excited by a specific light source. A plane is illuminated to observe the fluorescence of the particles. A camera then follows the movement of the dye dots, and the images are analyzed to characterize the flow. Depending on the photochromic dye, the reverse reaction to return to the nonfluorescent state varies from a few seconds to several minutes. Tomography, on the other hand, involves imaging objects by superimposing slices that are constructed with penetrating waves. In the case of electrical resistance, it is based on conductivity distribution images.

4.4.2. *Flow and velocity fields*

Historically, agitation-induced flows were demonstrated by simple observation, with or without dye, to determine whether the mixture was compatible with the intended application. Technologies have since evolved to record both global and local quantitative data on fluid velocity and flow (Mavros 2001). They can be divided into two categories: single-point measurement techniques, in other words, point and local, and ensemble measurement techniques, generally along a plane.

Local point measurement techniques are as follows:

– The *Pitot tube* (or Pitot probe) is based on measuring the pressure difference between the main flow and the flow after deflection towards an obstacle. Bernoulli's equation is then used to calculate fluid velocity. This technique has a long history (1724), but is intrusive and can affect flow, particularly in small tanks.

– *Hot-wire anemometry* consists of two probes connected by a taut wire (tungsten, platinum or platinum-iridium). An electric current is sent through the wire to heat it. An energy balance then calculates the cooling effect of the fluid on the wire, as a function of fluid velocity.

– *Laser Doppler velocimetry* (LDV) uses the intersection of two laser beams. A shift in wave frequency occurs if there is relative motion between the wave source and the wave receiver, the Doppler effect. The local velocity is determined using this frequency.

– *Ultrasonic Doppler velocimetry* (UDV) measures velocity using the Doppler frequency in a received signal. A transmitter sends out periodic ultrasonic waves, while a receiver continuously collects echoes of the waves reflected by inert particles moving with the fluid. The difference in frequency between transmitter and receiver is then used to calculate fluid velocity.

Overall measurement techniques are as follows:

– *Simple imaging* is based on visual determination of the flows generated by the fluid flow from still images or streak lines. Streak lines are possible with the addition of inert particles. Their movements are then analyzed using a laser plane.

– *Lagrangian ("flow-follower") tracking techniques*, such as positron emission particle tracking (PEPT) and computer automated radioactive tracking (CARPT), record the movement of specific particles to describe flow and circulation loops, or measure circulation time.

– *Color change* involves adding a color indicator (or two reagents to create a color indicator). This technique is limited to transparent fluids, but highlights flows and circulation loops. In some cases, a light source coupled with a camera can be used to record, process and reconstruct the mixing history.

– *Thermography* determines velocities using heat-sensitive thermochromic crystals injected into the fluid.

– *Particle image velocimetry* (PIV) involves analyzing two images of a plane passing through the tank, illuminated by a light source. The short time interval between the two images enables them to be compared, as well as the movement of each inert particle to be determined. Based on this image analysis and depending on the number of particles, flow maps can be constructed, along with the calculation of parameters characterizing the flow (e.g. vorticity and turbulence).

– The previously described pLIF for measuring mixing times can also be used, in a similar way to PIV, to reconstruct the movement of inert, and in this case, fluorescent, particles.

– The *tomography* technique also described above enables velocity fields to be established through the dispersion of inert particles. In this case, gamma rays, X-rays or ultrasonic waves are used to detect the dispersed phase. The main limitation of this technique is cost, and in the case of ultrasonic waves, only low solid load and gas retention are permitted (< 20%).

4.4.3. *Computational fluid dynamics simulation*

4.4.3.1. *General principle*

CFD is a digital approach to predicting flows, and the transfer of matter and heat, by solving equations and balances governing the motion of the fluids involved (Euler, Navier–Stokes, Boltzmann equations, etc.). The choice of the system of partial differential equations to be solved depends on the level of precision required for the physical representation of phenomena, and on the computing resources available. CFD complements the experimental approach by providing a better description of certain phenomena based on information that is more difficult to access experimentally (descriptive use) or by guiding the experimental approach, optimizing design sizing, or scaling up (predictive use).

Typically, simulation follows four main stages: (i) the preprocessing phase, including definition of the geometry and mesh; (ii) definition of boundary conditions and choice of models (set of equations to be solved); (iii) numerical resolution of the equations and (iv) post-processing of the results.

– The geometry is constructed using computer-aided design (CAD), which represents the zones occupied by the fluids whose hydrodynamics are to be simulated.

– A mesh is then prepared, allowing the transport equations to be discretized on it, so that they can be solved algebraically. The most common discretization techniques are the finite volume method (FVM) (around 80%), the finite element method (FEM) (around 15%) and the finite difference method (FDM) (around 5%) (Bakker 2006). FDM replaces partial derivatives with expansion series, often Taylor series. While this method is historical and simpler to code, it is limited to simple grids and runs the risk of poor conservation of balances on coarse grids. FEM integrates equations on an element or volume according to a weight function. Dependent variables are resolved by a shape function. Compared with FDM, it offers better accuracy on coarse grids and is particularly well suited to problems driven by diffusion, viscosity and free surfaces. However, it leads to slow resolutions for the largest problems to be solved, and it is unsuitable for turbulent flows. FVM integrates equations on a mesh volume. The dependent variables are solved using a linear variation approximation. This has the advantage of always preserving the conservation of matter, momentum and energy. In addition to these so-called classical methods, the recent technological boom in graphics processors has enabled higher performance calculations. This has made Boltzmann's lattice-based method more accessible and increasingly attractive, thanks to its high degree of parallelization (*lattice Boltzmann*). In this case, discretization is based on the collision and propagation of functions between meshes on a network that is usually cubic, and it defined by the number of domain dimensions and the number of propagation directions. There are, for example, applications using the D3Q19 network for velocity discretization, with three dimensions and 19 directions. For all discretization schemes, the number of meshes corresponds to the operator's choice between a reduced resolution time but a coarse resolution (a few hundred thousand meshes) and a more precise resolution but a longer calculation time (a few million to tens of millions of meshes, which can lead to calculations lasting several weeks or months in complex cases). It should be noted that the choice of models may require a sufficiently fine mesh to represent the physical phenomena being targeted.

– Models are parameterized and solved using codes that are developed in-house or using commercial software (ANSYS Fluent, CFX, OpenFoam, M-Star, etc.). Calculations are iterative until converged values are obtained. It is up to the modeler to define the convergence criteria.

– Lastly, post-processing is used to quantify and represent the physical quantities of interest at the end of the converged calculation. These enable the simulation objectives to be met (characterization, comparison, optimization, etc.).

4.4.3.2. *Interest for mixing in agitated tanks*

The discretization of the domain to solve fluid motion provides local information and meets the need to know volume distributions discussed in section 4.3.3. For example, the existence of a RTD dead zone or insufficient mixing time can be precisely identified in the tank, or even corrected, thanks to CFD characterization. CFD simulations can also be used to assess the risks associated with intense shearing at any point in the tank during mixing involving a sensitive compound.

In the case of turbulent flows, the size of the turbulent eddies varies from the size of the tank to the Kolmogorov scale. Thus, a simulation of the entire turbulent cascade requires a mesh resolution close to that of λ_k. In this case, a DNS (direct numerical simulation) is performed. However, this type of simulation requires mesh sizes in the range of a few tens of micrometers, increasing the computational resources required. A compromise is to increase the mesh size and partially model certain flow scales, where it is no longer possible to resolve all flow structures directly. The Taylor scale is defined as $\lambda_T = (10 \cdot \nu \cdot k/\varepsilon)^{1/2}$. Below this scale, microturbulence is deemed universal. The validity of sub-mesh models is therefore better. In the case of large-scale simulation, the LES (large eddy simulation) approach is used. Lastly, turbulence transport equations can be established by models, the most common of which are the k-ε and k-ω models. These various approaches make it possible to determine, be it directly or indirectly, the rate of dissipation of turbulent kinetic energy ε at any point in the fermenter, and de facto to determine the power dissipated within the mixed fluid. In the case of mixtures involving microorganisms, this local knowledge is all the more important as it enables the biological response (growth, cell death, production of molecules of interest, etc.) to be related to the environmental conditions actually encountered. It is thus possible to calculate the shear actually experienced by the biological phase, or by specifically filtering out the zones considered to be potentially damaging (e.g. by only considering eddies below a threshold diameter), and to adapt mixing conditions with the aim of improving process performance (Loubière et al. 2019).

4.5. Hydrodynamics–kinetics coupling

The development of numerical methods to reduce calculation times has led to a coupling between the mixture, characterized by a hydrodynamic model, and the biological response, characterized by a kinetic model. This chapter will mention biological response in relation to fermentation processes involving microorganisms or cell culture. This may involve growth, death of the biological phase, substrate consumption or the production of metabolites and molecules of interest.

4.5.1. *Kinetic models*

Many efforts have been made to better understand the behavior of microorganisms according to the environment in the bioreactor. However, even in a homogeneous environment, heterogeneity appears within the population, due to stochastic fluctuations on the scale of biochemical reactions (Delvigne and Goffin 2014). Microbial growth modeling is generally classified according to structured or unstructured, segregated or nonsegregated approaches (Nielsen et al. 2012), as summarized in Figure 4.13:

– In *unstructured models*, the cell population is treated as a single element. Biomass is described by a single variable, usually total biomass concentration (approximating balanced growth).

– In *structured models*, biomass is treated as a multi-component system, that is to say, the existence of intracellular elements is assumed. It is described by more than one variable.

– In *nonsegregated models*, cells are considered to be homogeneous. The model is then based on an averaged description of the cell.

– In *segregated models*, cells are considered heterogeneous, implying a population with cells of different properties. A population balance is then generally formulated. For example, in a given population, microorganisms of different ages may be considered.

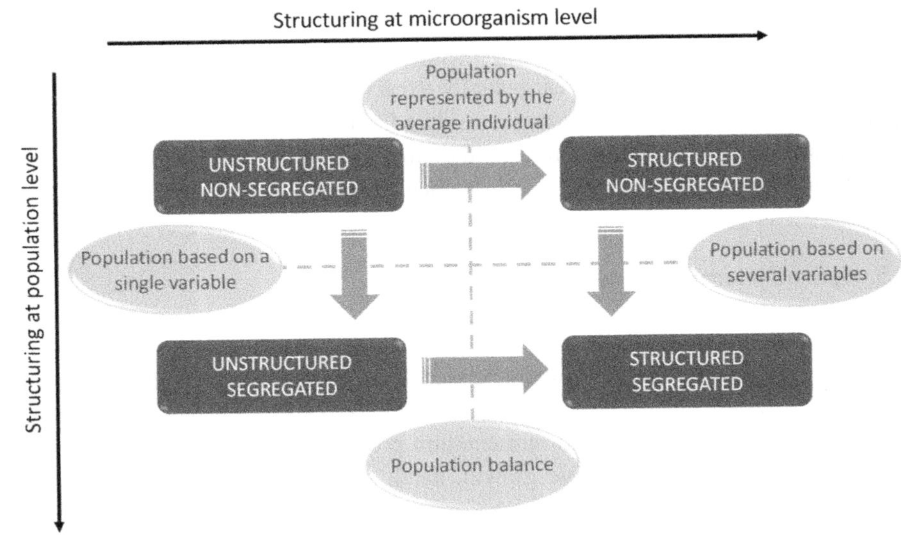

Figure 4.13. *Representation of the different types of biological models. For a color version of this figure, see www.iste.co.uk/ghoul/bioreactor.zip*

4.5.2. *Coupling issues*

Modeling hydrodynamics in agitated culture bioreactors can assume one to three phases if a solid phase or aeration is involved. However, cells or microorganisms can be considered as a fourth phase involving mass transfer. In general, bidirectional coupling can be used, where the liquid and biological phases influence one another.

The choice of the type of kinetic model generally depends on the degree of complexity. For example, a metabolic model with adaptation of the biological response to environmental variations will have a better predictive capability than one based on a constant conversion rate. It is therefore important to choose the model in conjunction with the desired spatial resolution of the hydrodynamic model, as well as the biological description within it. It would be inappropriate to couple an unstructured global kinetic model with a fine CFD model. The issues of interactions between mixing and biological response can be characterized by comparing physical time scales (micromixing, mesomixing and macromixing times) with bio-reactional time scales (characteristic consumption time τ_S) (Morchain 2017). Three situations can be distinguished:

– First, if the *macromixing* timescale is $\tau_{macro} \ll \tau_S$, then mixing is reached before the reaction can locally change the substrate concentration. The reactor is considered to be homogeneous and traditional approaches for ideal reactors can be applied. If $\tau_M \approx \tau_S$, then macroscopic gradients can be observed. A finer hydrodynamic model is recommended (e.g. compartmental approach in CFD), with a structured kinetic model to capture the impact of frequency and exposure of the biological phase to different concentrations.

– Then, if the *mesomixture* time scale is $\tau_{meso} \approx \tau_S$, then the reactor can appear homogeneous except near injection zones (which is the case for fed-batch and continuous reactors). In this case, the majority of the biological phase will be subjected to a concentration that is close to the average concentration, while a minority fraction of the biological phase will be exposed to very high concentrations (e.g. local injection of nutrients, acid and base). To capture such phenomena, a sub-grid model or spatial refinement can be applied to reflect local heterogeneity.

– Lastly, if the *micromixture* timescale is $\tau_{micro} \ll \tau_S$, then consumption is limited by cell-scale mass transfer, in other words, substrate diffusion across the cell membrane.

4.5.3. *Coupling applications*

4.5.3.1. *Unstructured and nonsegregated model*

Unstructured, nonsegregated models are based on the monitoring of the biological phase and nutrient concentrations, and they describe the process as an average cell population under ideal conditions. They are generally based on conservation equations for cell growth, nutrient and metabolic rates. Restricting ourselves to the exponential growth phase, the growth rate r_X (g·L^{-1}·h^{-1}) and substrate consumption rate r_S (g·L^{-1}·h^{-1}) can be expressed according to:

$$r'''_X = \frac{dX}{dt} = (\mu + k_D) \cdot X \quad [4.40]$$

$$r'''_S = -\mu \cdot \frac{X}{Y_{X/S}} \quad [4.41]$$

with X being the biomass concentration (g·L^{-1}), μ being the specific growth rate (h^{-1}), k_D being the mortality rate (often taken to be zero) and $Y_{X/S}$ being the substrate-to-biomass conversion efficiency. A model commonly used to predict μ as a direct function of nutrient concentration is the Monod equation (equation [4.42]), with μ_m being the maximum specific growth rate, S being the substrate concentration and K_S being the saturation constant.

$$\mu = \frac{\mu_m \cdot S}{K_S + S} \quad [4.42]$$

NOTE.– Due to resource and computing time limitations, early approaches coupling CFD and kinetic modeling were based on such "simple" kinetic models. However, they do not take population heterogeneity into account.

4.5.3.2. *Population balance*

The population balance model is a segregation model. It can be used to describe the biological heterogeneity induced by the gradients that are present in the reactor. These models have, for example, been implemented in CFD codes to simulate the dynamic response of microorganisms in the case of a piston reactor with concentration levels (Morchain et al. 2013). In this study, the biological response was set by the specific growth rate of the biomass, such that a single parameter governed segregation in the model, but opened the potential of the coupling-type approach on more parameters (multi-dimensional population balance).

4.5.3.3. *Lagrangian approach*

In so-called Euler–Lagrange approaches, the biological phase is simulated by virtual particles or parcel, where the state (set of properties) is "carried" by a vector-type internal parameter (Lapin et al. 2004). Such approaches enable the use of structured models, particularly metabolic models (Haringa et al. 2017). In addition, these approaches offer the possibility of tracking parcel lifelines, and thus expressing parameters as a function of time for each plot according to the intra-cellular response and extra-cellular environment "seen" by the biological phase. In other words, it is possible to capture the environmental fluctuations that microorganisms are subjected to over time.

The parameterization of the Lagrangian model (number of plots) strongly depends on how consumption reactions are controlled by the biological phase. Haringa et al. (2018) distinguish three situations as follows:

– *No fast or slow intracellular control*: this corresponds to an instantaneous intracellular response or a constant intracellular environment (e.g. simple Monod kinetics with fixed constants). The population is homogeneous, both microbiologically and spatially.

– *Slow intracellular control only*: the population is heterogeneous from a microbial point of view, but remains spatially homogeneous due to a much lower mixing time compared to pool variations ($\tau_{pool} \gg \tau_{mix}$).

– *Fast control only or fast and slow control*: local consumption reactions are affected by intra-cellular heterogeneity in rapid pool turnover. This means that concentration fluctuations experienced by microorganisms' condition bioreactor dynamics.

4.6. Conclusion

Agitated bioreactors provide a suitable environment for the cultivation of microorganisms for biomass expansion, production of molecules of interest, bioconversion, etc.

Their characterization on a macroscopic scale makes it possible to target designs and operating conditions adapted to mixing objectives. Various parameters that can be measured or calculated according to their formulas or from semi-empirical correlations enable flow, particle suspension and aeration regimes to be identified, as well as characteristic quantities such as power required or mixing time. These elements constitute a first engineering approach for defining the operating conditions and agitation design that meet the process mixing objectives.

In processes such as fermentation, it may be necessary to understand phenomena on a microscopic scale in order to consider the impact of hydromechanical stresses on microorganisms. This involves estimating local quantities such as eddy size distribution in a turbulent regime.

These characterizations are the subject of ongoing development, both through experimental and modeling approaches. The succession of microenvironments undergone by microorganisms and their heterogeneities are increasingly considered in the modeling of biological response, opening up new challenges for understanding and improving process performance.

4.7. References

Arvinth, S., Rao, P.G., Murugesan, T. (1996). Critical impeller speed for solid suspension in turbine agitated contactors. *Bioprocess and Biosystems Engineering*, 14, 97–99.

Ascanio, G. (2015). Mixing time in stirred vessels: A review of experimental techniques. *Chinese Journal of Chemical Engineering*, 23(7), 1065–1076.

Bakker, A. and Marshall, E.M. (2006). Computational fluid dynamics. Lecture 4 – Classification of Flows [Online]. Available at: https://www.bakker.org/Lectures-Applied-CFD.pdf.

Baldi, G., Conti, R., Alaria, E. (1978). Complete suspension of particles in mechanically agitated vessels. *Chemical Engineering Science*, 33(1), 21–25.

Calderbank, P.H. (1958). Physical rate processes in industrial fermentation. Part 1: The interfacial area in gas-liquid contacting with mechanical agitation. *Trans. Inst. Chem. Eng.*, 36, 443–463.

Chapple, D., Kresta, S.M., Wall, A., Afacan, A. (2002). The effect of impeller and tank geometry on power number for a pitched blade turbine. *Chemical Engineering Research and Des*ign, 80(4), 364–372.

Danckwerts, P.V. (1958). The effect of incomplete mixing on homogeneous reactions. *Chemical Engineering Science*, 8(1–2), 93–102.

Delafosse, A., Loubière, C., Calvo, S., Toye, D., Olmos, E. (2018). Solid-liquid suspension of microcarriers in stirred tank bioreactor – Experimental and numerical analysis. *Chemical Engineering Science*, 180, 52–63.

Delvigne, F. and Goffin, P. (2014). Microbial heterogeneity affects bioprocess robustness: Dynamic single-cell analysis contributes to understanding of microbial populations. *Biotechnology Journal*, 9(1), 61–72.

Dew, D.W., Rorke, G.V., Guezennec, A.G. (2022). Engineering designs and challenges of stirred tank systems. In *Biomining Technologies: Extracting and Recovering Metals from Ores and Wastes*. Springer, Cham.

Haringa, C., Noorman, H.J., Mudde, R.F. (2017). Lagrangian modeling of hydrodynamic–kinetic interactions in (bio)chemical reactors: Practical implementation and setup guidelines. *Chemical Engineering Science*, 157, 159–168.

Haringa, C., Mudde, R.F., Noorman, H.J. (2018). From industrial fermentor to CFD-guided downscaling: What have we learned? *Biochemical Engineering Journal*, 140, 57–71. doi: 10.1016/j.bej.2018.09.001.

Henzler, H.-J. (2000). Particle stress in bioreactors. In *Influence of Stress on Cell Growth and Product Formation*, Schügerl, K., Kretzmer, G., Henzler, H., Kieran, P., Kretzmer, G., MacLoughlin, P., Malone, D.M., Schuman, W., Shamlou, P.A., Yim, S.S. (eds). Springer, Berlin, Heidelberg.

Kee, N.C. and Tan, R.B. (2002). CFD simulation of solids suspension in mixing vessels. *The Canadian Journal of Chemical Engineering*, 80(4), 1–6.

Kolmogorov, A.N. (1941). Dissipation of energy in the locally isotropic turbulence. *Dokl. Akad. Nauk. SSSR*, 32, 19–21.

Kolmogorov, A.N. (1991). The local structure of turbulence in incompressible viscous fluid for very large Reynolds numbers. *Proceedings of the Royal Society of London. Series A: Mathematical and Physical Sciences*, 434(1890), 9–13.

Kraume, M. (1992). Mixing times in stirred suspensions. *Chemical Engineering & Technology: Industrial Chemistry-Plant Equipment-Process Engineering-Biotechnology*, 15(5), 313–318.

Kresta, S.M., Etchells III, A.W., Dickey, D.S., Atiemo-Obeng, V.A. (eds) (2015). *Advances in Industrial Mixing: A Companion to the Handbook of Industrial Mixing*. John Wiley & Sons, Hoboken.

Lapin, A., Müller, D., Reuss, M. (2004). Dynamic behavior of microbial populations in stirred bioreactors simulated with Euler–Lagrange methods: Traveling along the lifelines of single cells. *Industrial & Engineering Chemistry Research*, 43(16), 4647–4656.

Loubière, C., Delafosse, A., Guedon, E., Toye, D., Chevalot, I., Olmos, E. (2019). Optimization of the impeller design for mesenchymal stem cell culture on microcarriers in bioreactors. *Chemical Engineering & Technology*, 42(8), 1702–1708.

Mak, A.T.C. (1992). *Solid-Liquid Mixing in Mechanically Agitated Vessels*. University College London, London.

Mavros, P. (2001). Flow visualization in stirred vessels: A review of experimental techniques. *Chemical Engineering Research and Design*, 79(2), 113–127.

Mersmann, A., Werner, F., Maurer, S., Bartosch, K. (1998). Theoretical prediction of the minimum stirrer speed in mechanically agitated suspensions. *Chemical Engineering and Processing: Process Intensification*, 37(6), 503–510.

Morchain, J. (2017). Numerical tools for scaling up bioreactors. In *Current Developments in Biotechnology and Bioengineering*, Larroche, C., Sanromán, M.A., Du, G., Pandey, A. (eds). Elsevier, Amsterdam.

Morchain, J., Gabelle, J.-C., Cockx, A. (2013). Coupling of biokinetic and population balance models to account for biological heterogeneity in bioreactors. *AIChE Journal*, 59(2), 369–379.

Nielsen, J., Villadsen, J., Lidén, G. (2012). *Bioreaction Engineering Principles: Second Edition*. Springer, Boston.

Nienow, A. (1968). Suspension of solid particles in turbine agitated baffled vessels. *Chemical Engineering Science*, 23, 1453–1459.

Paul, E.L., Atiemo-Obeng, V.A., Kresta, S.M. (2004). *Handbook of Industrial Mixing*. Wiley-Blackwell. Hoboken.

Raghava Rao, K.S.M.S., Rewatkar, V.B., Joshi, J.B. (1988). Critical impeller speed for solid suspension in mechanically agitated contactors. *AIChE Journal*, 34(8), 1332–1340.

Roustan, M. (1997). Agitation. Mélange : caractéristiques des mobiles d'agitation. *Techniques de l'Ingénieur. Génie des procédés*, 4, J3802-1.

Tamburini, A., Cipollina, A., Micale, G., Brucato, A., Ciofalo, M. (2011). CFD simulations of dense solid–liquid suspensions in baffled stirred tanks: Prediction of suspension curves. *Chemical Engineering Journal*, 178, 324–341.

Tamburini, A., Brucato, A., Cipollina, A., Micale, G., Ciofalo, M. (2012). CFD predictions of sufficient suspension conditions in solid-liquid agitated tanks. *International Journal of Nonlinear Sciences and Numerical Simulation*, 13, 427–443.

Villermaux, J. (1993). *Génie de la réaction chimique. Conception et fonctionnement des réacteurs*. Lavoisier, Paris.

Wichterle, K. (1988). Conditions for suspension of solids in agitated vessels. *Chemical Engineering Science*, 43(3), 467–471.

You, S.T., Raman, A.A.A., Shah, R.S.S.R.E., Mohamad Nor, M.I. (2014). Multiple-impeller stirred vessel studies. *Reviews in Chemical Engineering*, 30(3), 323–336.

Zhou, G. and Kresta, S.M. (1996). Impact of tank geometry on the maximum turbulence energy dissipation rate for impellers. *AIChE Journal*, 42(9), 2476–2490.

Zwietering, T.N. (1958). Suspending of solid particles in liquid by agitators. *Chemical Engineering Science*, 8, 244–253.

Zwietering, T.N. (1959). The degree of mixing in continuous flow systems. *Chemical Engineering Science*, 11(1), 1–15.

5

Bioreactor Extrapolation

Céline LOUBIÈRE, Eric OLMOS and Mohamed GHOUL
Laboratoire Réaction et Génie des Procédés (LRGP), CNRS, Université de Lorraine, Nancy, France

5.1. Introduction

Mastering bioreactor extrapolation techniques is an essential step in achieving the efficiency and productivity required to ensure the economic profitability of the products sought. Several criteria need to be considered, depending on the nature of the bacterial strain used, the target product, the bioreactor configuration, etc. It is therefore necessary to find compromises between these criteria.

5.2. Introducing the bioreactor extrapolation concept

Extrapolation is a methodological approach aimed at improving production when changing production scale, whether it be from laboratory to pilot or industrial scale, using the same reagents and maintaining similar critical process parameters in the bioreactor. The general idea is to characterize and identify operating conditions that minimize the impact of the scale-up (Schmidt 2005).

Historically, the first case study of scale-up took place shortly after the Second World War to meet the need for mass production of penicillin. The application of a methodology to control scale-up then became critical in the process development, improving product quality while reducing production costs (Hourcade 1985). It has also made it possible to overcome certain shortcomings observed in small-scale

Bioreactor Implementation in the Agro-Food Industries,
coordinated by Mohamed GHOUL.

production systems, such as difficulties in regulating pH and dissolved oxygen in non-monitored systems, which can lead to poor yields.

There are two types of approach, depending on the objective:

– The "*scale-up*" approach is applied when a process is taken from a small scale to a larger scale, such as from laboratory to pilot or industrial scale. Critical factors influencing process performance are identified at the small scale and optimized, before defining the operating conditions at the large scale, which will enable these conditions to be maintained. This approach is particularly useful when scaling up new processes.

– The "*scale-down*" approach relies on downscaling, so as to improve process performance. It is generally applied to existing processes, where the aim is to reproduce the environment observed at the industrial scale, but at the laboratory scale. A good knowledge of the industrial process is a prerequisite, in particular of the heterogeneities that may only appear at a certain scale. The operating conditions influencing performance are then identified and, if necessary, improved before being applied on an industrial scale (Möckel et al. 1990).

In food and biotechnology bioprocesses, the implementation of a new process requires the selection of strain, culture medium composition, bioreactor design and operating conditions (agitation, aeration, culture mode, etc.). These choices impact production performance, may or may not be coupled, and must be optimized. This step enables the technical performance of the process to be assessed, and the hydrodynamics, thermodynamics and kinetic data required for scaling up to be obtained. Due to the complex nature of the biological phase, the development of bioprocesses requires considerable time and capital investment. A number of strategies have therefore been developed in the early stages of development at the smallest culture scales in order to have sufficiently robust data to prepare for the scale-up operation: statistical experimental design, preliminary developments in fed-batch mode, screening in mini-bioreactors, use of online sensors for monitoring and control tools in combination with automation of manual stages, process modeling, etc. (Neubauer et al. 2013). Taking critical parameters into account as early as possible in process development facilitates subsequent scale-up.

The transition from the laboratory to the industrial scale is achieved progressively in several stages in order to avoid abrupt variations in system hydrodynamics. A ratio of 10 (v/v) is often applied at each scale-up stage. Initially, process characterization involves analysis of physicochemical parameters (maximum real-time data in situ, with possible chemometric analysis and modeling) and mixing parameters, taking into account the mutual influences of physicochemical and physical parameters. This stage is often jointly based on experimental data that can be supported by statistical processing, such as neural networks (empirical approach), and on theoretical hypotheses and models (mechanistic approach). Subsequently, process optimization involves process control and relevant design (scale-down approach, CFD, etc.) and aims to improve physiological reaction conditions and reduce exposure to stress, at the given scale. This stage also enables us to identify the appropriate scale-up strategy for subsequent stages.

The gradual transition from the laboratory, pilot and industrial stages involves a number of similarity laws, which help to maintain constant criteria that is specific to the system under study (Ghasem 2007). The most commonly used similarity laws for mixing in bioreactors are:

– *geometric similarity*: conservation of the same scale ratio;

– *reaction similarity*: conservation of growth kinetics, substrate consumption, metabolite production, etc.;

– *thermal similarity*: conservation of the same temperature gradient.

We should also mention the dynamics (conservation of forces at all points) and kinematics (conservation of velocities at all points) similarities that exist in mechanics, but which are less applied in the case of mixing issues in agitated tanks. The application of these similarity laws is based on the inter-scale conservation of several characteristic, dimensionless numbers, enabling independence of scale, such as the Reynolds number (Re), the Froude number (Fr), the Weber number (We) or characteristic parameters like volumetric power (P/V), impeller tip speed (Ntip), mixing time (tm), oxygen transfer coefficient (KLa), etc. (Junker 2004).

NOTE.– Scale-up must address hydrodynamics, possibly in conjunction with mass transfer when aeration is applied. It is based on detailed process characterization and timely identification of key process parameters that would affect product efficiency, quality and conformity (Marques et al. 2010). However, bear in mind that any approach and model will always be approximate and will represent compromises. The mathematical methods and relationships used to establish scale-up models cannot fully model the complexity of interactions and relationships between the physical parameters that govern the process (Lidén 2002). As a result, scaling, in the classical sense of the term, can be likened to a “black box” method of representing cultivation conditions, which does not necessarily consider local phenomena in the bioreactor, and characterizes it by macroscopic quantities.

5.3. Extrapolation methods

Two extrapolation techniques are generally applied. The first, known as "geometric", is used in cases where it is possible to maintain a constant ratio between dimensions at the different stages of process development (laboratory, pilot and industrial). The second, known as "non-geometric", is applied when at least one of the dimensional relationships between the scales is different.

In practice, depending on the situation, it is determined whether or not geometric similarity can be applied. Next, the operating conditions are identified, so as to keep one or more key process parameters constant. The expert will choose these parameters according to the process and its constraints, being aware that not all parameters can necessarily be maintained simultaneously by similarity. In bioprocesses, agitation and aeration are two key operations. They can be optimized either as a result of geometric similarity, or for already existing and installed configurations, without considering geometric similarity criteria. In both cases, the focus is generally on agitator geometry, and scale-up can then be carried out on criteria involving agitation, aeration or a combination of the two. The most frequently encountered cases in fermentation are mixture maintenance based on P/V (non-aerated condition) and aeration maintenance based on volume oxygen transfer rate (OTR) or k_La. After presenting the so-called geometric and non-geometric extrapolations, these two scaling strategies will be discussed in more detail in this chapter. The first case concerns extrapolation under single-phase conditions (or with a dilute dispersion of solids and physical properties that do not significantly affect the flow), before tackling the case of a mixture involving gas injection.

5.3.1. *Geometric and non-geometric similarities*

Most extrapolation in food bioprocesses (production of alcohol, single-cell proteins or organic acids) use geometric extrapolation methodology (Junker 2004; Ghasem 2007), retaining certain dimensionless quantities that are characteristic of the system studied, as well as the physicochemical and biological parameters.

The most commonly used parameters and dimensionless numbers are shown in Table 5.1. These include Reynolds, Froude and Weber numbers, as well as impeller tip speed, rotation frequency, mixing time, volumetric power, pH and dissolved oxygen profiles (Nielsen et al. 2003; Marques et al. 2010). From a practical point of view, the application of an extrapolation process requires prior identification of all factors likely to cause variations in the selected dimensionless criteria. Among the most important factors are temperature, pressure, composition and viscosity of the medium, bioreactor configuration and the strain used (Nienow 1998).

Scaling criteria	Symbols	Formulas	Proportionality
Reynolds number	Re	$\frac{\rho \cdot N \cdot D^2}{\mu}$	$Re \propto N \cdot D^2$
Froude number	Fr	$\frac{N^2 \cdot D}{g}$	$Re \propto N^2 \cdot D$
Weber number	We	$\frac{\rho \cdot N^2 \cdot D^3}{\sigma}$	$We \propto N^2 \cdot D^3$
Volumetric power (non-aerated condition)	P/V	$\frac{Np \cdot \rho \cdot N^3 \cdot D^5}{V}$	$P/V \propto N^3 \cdot D^2$
Volumetric power (aerated condition)	P_g/V	$f(P, N_A)$	
Impeller tip speed	u_{tip}	$\pi \cdot N \cdot D$	$u_{tip} \propto N \cdot D$
Mixing time	t_m	$t_m = \frac{V}{Q_p} = \frac{V}{N_{Q_p} N D^3}$	$t_m \propto N^{-1}$
Dimensionless mixing time	$t_m N$	$t_m N = f(Re, N_p)$	$t_m N \propto 1$
Modified dimensionless mixing time	$t_m'N$	$t_m'N = \frac{N t_m}{\left(\frac{T}{D}\right)^2 Re} = \frac{t_m \mu}{T^2 \rho}$	
Aeration rate	A_R	$A_R = \frac{Q_g}{V}$	
Surface gas velocity	u_g	$u_g = \frac{4Q_g}{\pi T^2}$	
Gas retention	τ_g	$\tau_g = \frac{V}{Q_g}$	
Gas retention rate	ε_g	$\varepsilon_g = \frac{V_g}{V}$	
Aeration number	N_A	$N_A = \frac{Q_g}{N D^3}$	
Volumetric oxygen transfer coefficient	$k_L a$	$k_L a = a' \left(\frac{P}{V}\right)^b u_g^c$	

ρ: density of medium, N: agitation frequency (rps), D: agitation diameter (m), μ: dynamic viscosity (Pa·s), g: gravitational constant (($m^3 \cdot kg^{-1} \cdot s^{-2}$), N_p: power number (-), σ: surface tension ($N \cdot m^{-1}$), Q_P: pumping flow rate ($m^3 \cdot s^{-1}$), N_{Qp}: pumping number (-), V: Volume (m^3), T: tank diameter (m), P: Power (W), σ: interfacial tension ($N \cdot m^{-1}$), μ: dynamic viscosity (Pa·s), Q_g: gas flow rate ($m^3 \cdot s^{-1}$).

Table 5.1. *Main dimensionless numbers and parameters used in bioreactor extrapolation*

Let us consider a geometric similarity based on geometric ratios, where indices 1 and 2 represent the two distinct scales of study, for example, industrial and laboratory. We then define the scaling factor F according to equation [5.1].

$$F = \frac{D_2}{D_1} = \frac{T_2}{T_1} = \frac{H_2}{H_1} = \frac{C_2}{C_1} \quad [5.1]$$

This scaling factor can then be used to establish the relationship between maintaining a given criterion and other process parameters.

In some cases, it is not possible to comply with the geometric similarities between scales, as the dimensions of the bioreactor are already fixed and only the agitation impeller can be optimized. This situation may arise when the biocatalyst used (microorganisms or enzymes) is sensitive to shear forces. In this case, extrapolation focuses on the D/T ratio (agitation diameter to bioreactor height) or agitator height. This is known as partial geometric similarity.

As with geometric extrapolation, if we consider that the agitator is characterized by its diameter D, the criteria based on the Reynolds number, power per unit volume, impeller tip speed, as well as the Froude and Weber numbers, allow proportionality as a function of N and D, as presented in Table 5.1.

5.3.2. *Scaling in non-aerated, agitation-based conditions*

Agitation, through mixing and power consumption, is a key parameter from an engineering perspective. This power can vary between 10 $W{\cdot}m^{-3}$ (in the case of fragile or viscous systems) and 3 $kW{\cdot}m^{-3}$ (in the case of systems with high oxygen demand). Mixing objectives can guide extrapolation on these parameters when, for example, maintaining a low shear rate in the presence of stress-sensitive microorganisms is essential. Catalyst shear strength is naturally the most important criterion. Various criteria can be set for extrapolation, including impeller tip speed (utip), the Reynolds number (*Re*) and volumetric power (*P/V*).

5.3.2.1. *Extrapolation based on impeller tip speed*

For a long time, due to its ease of implementation, studies have been based on maintaining impeller tip speed (Hewitt and Nienow 2007; Marques et al. 2010). However, from a hydrodynamic point of view, the velocity gradient in the laminar regime should be estimated from the quotient of the impeller tip speed and the space between the impeller and the vessel wall, as represented in equation [5.2].

This equation shows that by keeping the impeller tip speed constant during extrapolation, the T-D distance is not conserved, nor is the shear.

$$\dot{\gamma} = \frac{\pi \cdot N \cdot D}{(T-D)/2} \qquad [5.2]$$

Furthermore, in the turbulent regime, the shear rate is provided by $\dot{\gamma} = \sqrt{\varepsilon/\upsilon} = \sqrt{(P/V)/\mu}$. Knowing that *P/V* is proportional to N^3D^2, the shear rate is therefore proportional to $N^{1,5}D$, which differs from the *ND* = constant scaling law.

Ultimately, considering an extrapolation based on geometric similarity, an impeller tip speed maintained according to equation [5.3] leads to equation [5.4].

$$u_{tip} = \pi \cdot N_1 \cdot D_1 = \pi \cdot N_2 \cdot D_2 \qquad [5.3]$$

$$N_2 = N_1 \cdot \frac{D_1}{D_2} = N_1 \cdot \frac{1}{F} \qquad [5.4]$$

The expression of the power dissipated in the turbulent regime from the power number allows us to establish equations [5.5] and [5.6].

$$P_1 = N_p \cdot \rho \cdot N_1^3 \cdot D_1^5 \qquad [5.5]$$

$$P_2 = N_p \cdot \rho \cdot N_2^3 \cdot D_2^5 = N_p \cdot \rho \cdot \left(N_1 \cdot \frac{D_1}{D_2}\right)^3 \cdot D_2^5 = P1 \cdot \left(\frac{D_2}{D_1}\right)^2 \qquad [5.6]$$

From the scaling factor defined in equation [5.1], we can ultimately prove equation [5.7].

$$\left(\frac{P}{V}\right)_2 = \frac{1}{F} \cdot \left(\frac{P}{V}\right)_1 \qquad [5.7]$$

Extrapolation based on the maintenance of impeller tip speed therefore leads to a reduction in agitation frequency and volumetric power during scale-up. This has a significant impact on processes requiring aeration, with the risk of limiting oxygen transfer phenomena.

5.3.2.2. *Extrapolation based on volumetric power*

Maintaining the volumetric power *P/V* has gradually replaced the criterion based on impeller tip speed and has become the most widely applied agitation-based extrapolation parameter today.

Considering an extrapolation based on geometric similarity, maintaining the volumetric power, reduced to equation [5.8], leads to equation [5.9].

$$\left(\frac{P}{V}\right) \propto \frac{P}{T^3} \quad [5.8]$$

$$\frac{P_2}{T_2^3} = \frac{P_1}{T_1^3} \quad [5.9]$$

From the scaling factor defined in equation [5.1], we can ultimately prove equation [5.10].

$$P_2 = P_1 \cdot F^3 \quad [5.10]$$

This extrapolation criterion enables average mechanical stresses and material transfer capacities to be maintained, with a reasonable increase in mixing times.

Nevertheless, it requires information on volumetric dissipations, and leads to a sharp increase in total power dissipation, impacting operating costs.

5.3.2.3. *Extrapolation based on the Reynolds number*

By maintaining the Reynolds number *Re*, the flow regime is preserved.

Considering an extrapolation based on geometric similarity, maintaining the Reynolds number leads to equation [5.11].

$$Re_1 = \frac{\rho \cdot N_1 \cdot D_1^2}{\mu} = Re_2 = \frac{\rho \cdot N_2 \cdot D_2^2}{\mu} \quad [5.11]$$

From the scaling factor defined in equation [5.1], we can ultimately prove equations [5.12] and [5.13].

$$N_2 = N_1 \cdot \frac{1}{F^2} \quad [5.12]$$

$$P_2 = P_1 \cdot \frac{1}{F} \quad [5.13]$$

In this case, we demonstrate that the choice of this criterion implies a reduction in agitation frequency and power dissipation.

Maintaining the Reynolds number at least ensures that the flow regime is maintained during scale-up, and is of particular interest during scale-down operations, where there is a risk of losing the turbulent regime with the reduction in D. However, during scale-up, this criterion is not favorable for processes requiring good gas transfer, due to the reduction in agitation frequency and power dissipation. There is also a sharp increase in mixing times.

5.3.2.4. *Other extrapolation criteria based on agitation*

Other criteria are also sometimes used, such as mixing time (Grenville and Nienow 2004; Rodgers et al. 2011). This can be estimated from correlations such as equation [5.14] (Nienow 2009). The mixing time is compared with the typical times of chemical and biological reactions. When extrapolating, we ensure that t_m remains below these typical times.

$$t_m = 5.9 \cdot T^{\frac{2}{3}} \cdot \varepsilon^{-\frac{1}{3}} \cdot \left(\frac{D}{T}\right)^{-\frac{1}{3}} \qquad [5.14]$$

Another option is to maintain the agitation frequency N, so as to ensure constant pumping and mixing times, although this is not recommended. This is because the power dissipated during scale-up increases sharply, which is not recommended for most bioprocesses, or becomes incompatible with the technical means available.

Lastly, more recently, we find the criterion based on a dissipation and circulation energy function EDCF ($W \cdot m^{-3} \cdot s^{-1}$), defined equation [5.15] (Smith et al. 1990), with ε_{max} being the maximum turbulent kinetic energy dissipation rate and t_c being the circulation time (s). This parameter makes it possible to correlate agitation with the circulation rate of microorganisms near the agitation, where stress is highest.

$$\text{EDCF} = (\rho \cdot \varepsilon_{max}) \cdot \frac{1}{t_c} \qquad [5.15]$$

5.3.2.5. *Summary of the impact of scaling on process parameters*

Depending on the extrapolation criterion selected and held constant, the impact of the scaling on other criteria or process parameters can be estimated according to the order of magnitude applied to F, as previously demonstrated for maintaining impeller tip speed, volumetric power and the Reynolds number. Table 5.2 summarizes these effects for fluids maintaining the same viscosities and densities between scales (Roustan et al. 1999), and presents the effects on a wider set of possible criteria for extrapolation.

Variables	N (s^{-1})	V_{tip} ($m \cdot s^{-1}$)	P/V ($W \cdot m^{-3}$)	t_m (s)	Re (-)	Fr (-)
Dimensions	N	ND	N^3D^2	D	ND^2	N^2D
N (s^{-1})	$F^0 = 1$	F^{-1}	$F^{-0.67}$	$F^0 = 1$	F^{-2}	$F^{-0.5}$
$\dot{\gamma}_{moyen}$ (s^{-1})	$F^0 = 1$	F^{-1}	$F^{-0.67}$	$F^0 = 1$	F^{-2}	$F^{-0.5}$
u_{tip} ($m \cdot s^{-1}$)	F^1	$F^0 = 1$	$F^{0.33}$	F^1	F^{-1}	$F^{0.5}$
P (W)	F^5	F^2	F^3	F^5	F^{-1}	$F^{3.5}$
P/V ($W \cdot m^{-3}$)	F^2	F^{-1}	$F^0 = 1$	F^2	F^{-4}	$F^{0.5}$
t_p (s)	F^0	F^1	$F^{0.67}$	$F^0 = 1$	F^2	$F^{0.5}$
t_m (s)	F^0	F^1	$F^{0.67}$	$F^0 = 1$	F^2	$F^{0.5}$
Re (-)	F^2	F^1	$F^{1.33}$	F^2	$F^0 = 1$	$F^{1.5}$
Fr (-)	F^1	F^{-1}	$F^{-0.33}$	F^1	F^{-3}	$F^0 = 1$
H^* (m)	F^2	$F^0 = 1$	$F^{0.67}$	F^2	F^{-2}	F^1
Q_p ($m^3 \cdot s^{-1}$)	F^3	F^2	$F^{2.33}$	F^3	F^1	$F^{2.5}$
Q_p/H^* ($m^2 \cdot s^{-1}$)	F^1	F^2	$F^{1.67}$	F^1	F^3	$F^{1.5}$

N: agitation frequency, γ: velocity gradient, V_{tip}: impeller tip speed, P: power, V: volume, t_p: pumping time, t_m: mixing time, Re: Reynolds number, Fr: Froude number, H^*: theoretical height created by the agitation impeller, Qp: pumping flow rate. *Note*: parameters considered under turbulent conditions.

Table 5.2. *Orders of magnitude of the impact of scaling on process parameters in geometric similarity and under non-aerated conditions (constant density and viscosity)*

To illustrate the table, in the case of fluids with the same density and viscosity, and in the absence of aeration, keeping the rotation frequency, N, constant between the two scales ($N1 = N2$), we obtain the following effects:

– Effect on Reynolds number:

$$Re \propto N_2 \cdot D_2^2 \propto N_1 \cdot D_1^2 \cdot F^2 \qquad [5.16]$$

– Effect on volumetric power:

$$P/V \propto N_2^3 \cdot D_2^2 \propto N_1^3 \cdot D_1^2 \cdot F^2 \qquad [5.17]$$

– Effect on Froude number:

$$Fr \propto N_2^2 \cdot D_2 \propto N_1^2 \cdot D_1 \cdot F^1 \qquad [5.18]$$

– Effect on impeller tip speed:

$$V_{tip} \propto N_2 \cdot D_2 \propto N_1 \cdot D_1 \cdot F^1 \qquad [5.19]$$

Under these conditions, we can see that the impeller tip speed and Froude number will be influenced by F^1, the volumetric power and Reynolds number by F^2, and the Weber number by F^3. It is also worth noting that the setting of certain criteria leads to a significant variation in other variables, such as shear force or energy consumption. It is therefore essential to select the most important criteria to be considered, depending on the system under study.

5.3.3. *Scaling in aerated condition and based on aeration*

Most of the metabolites used in the food industry (flavors, colorants, biomass, etc.) are produced aerobically. Extrapolation must therefore be carried out while ensuring sufficient gas transfer in the fermenter. To achieve this, various criteria can be maintained, such as the flow rate expressed in vvm (volumetric flow of gas per liquid volume per minute), the oxygen transfer coefficient (k_La), the volumetric OTR or the aeration regime via the aeration number (Möckel et al. 1990).

This section will focus on the volumetric OTR via the oxygen transfer coefficient, which is used in most cases (Shukla et al. 2001; Galaction et al. 2004; Hewitt and Nienow 2007; Garcia-Ochoa and Gomez 2009; Marques et al. 2010).

5.3.3.1. *Representing the issue of oxygenation extrapolation*

Oxygen transfer in an agitated and aerated bioreactor can be broken down into a series of stages (transfer from gas to gas-liquid interface, transfer from interface to liquid phase, then convective transport in the liquid phase). The limiting stage in the dissolution process is generally considered to be diffusive transfer across the liquid boundary layer, close to the gas–liquid interface. The dissolved OTR ($mol \cdot L^{-1} \cdot h^{-1}$) is defined according to equation [5.20], with k_La being the oxygen transfer volume coefficient (h^{-1}), O_2 being the oxygen concentration in the liquid phase ($mol \cdot L^{-1}$) and O_2^* being the dissolved oxygen concentration when equilibrium with the oxygen partial pressure in the gas phase is reached (saturation threshold) ($mol \cdot L^{-1}$). O_2^* can be calculated from Henry's law, as shown in equation [5.21], with KH being Henry's constant ($mol \cdot L^{-1} \cdot Pa^{-1}$) and pO_2 being the oxygen partial pressure (Pa), which is expressed as a function of the oxygen fraction xO_2 and the total pressure P^{tot}. The operating conditions of aeration and agitation are chosen so that the OTR is sufficient to meet the oxygen demand of the process, which, in this case, is of the microorganisms.

$$OTR = k_L a \cdot (O_2^* - O_2) \quad [5.20]$$

$$O_2^* = K_H \cdot p_{O_2} = K_H \cdot x_{O_2} \cdot P^{tot} \quad [5.21]$$

In the context of scaling up, we will be looking for a process that maintains similar biological performance and therefore similar oxygen demand. While oxygen consumption is a controlled variable, OTR is adjustable and will be the criterion we aim to maintain, in order to run the process properly. Equation [5.20] shows that if, for example, we wish to increase oxygen dissolution capacity, we can either improve k_La or enrich the injected gas with oxygen, thus increasing O_2^*. An increase in the total pressure in the bioreactor also enhances the driving force for transfer $(O_2^* - O_2)$.

In Chapter 4, it was shown that the volumetric oxygen transfer coefficient is a function of the dissipated volumetric power under aerated conditions and the surface gas velocity (Chapter 4, section 4.2.5). So, considering the conservation of volumetric power under aerated conditions, a second extrapolation criterion must be considered in order to jointly determine the agitation frequency and gas volumetric flow rate to be applied during extrapolation.

We recall the Van't Riet equation and its variants, as shown in equations [5.22] and [5.23], with C_X being the cell density, and examples of resolutions on the culture database in Table 5.3.

$$k_L a = A \cdot \left(\frac{P_g}{V}\right)^{\alpha} \cdot u_g^{\beta} \quad [5.22]$$

$$k_L a = A \cdot \left(\frac{P_g}{V}\right)^{\alpha} \cdot u_g^{\beta} \cdot \frac{1}{C_X^{\gamma}} \quad [5.23]$$

The following sections present examples of extrapolation strategies, based on maintaining the selection of OTR and P_g/V, so as to determine the operating conditions to be applied following the scaling. They distinguish between situations and relative assumptions, where the pressures within the tanks at both scales allow a similar O_2^* concentration to be approximated, which is applicable to small scales, and where it is necessary to consider the deviation during the change of scale, amplified with the increase in scale.

Culture condition in stirred reactor	*A*	*α*	*B*	*γ*	References
With coalescence	0.002	0.7	0.2	–	Nielsen et al. 2003
Without coalescence	0.32	–	0.7	–	Nielsen et al. 2003
Laboratory scale (100 mL), *E. coli*	0.224	0.35	0.52	–	Gill et al. 2008
Laboratory scale (8 L)	Strain function	0.95	0.67	–	Junker 1998
Pilot scale (400 L)	Strain function	0.67	0.67	–	Junker 1998
Industrial scale (23,000–46,000 L)	Strain function	0.5	0.5	–	Junker 1998
Laboratory scale (4 L), *P. shermani*	6.586	–0.0286	0.429	0.282	Galaction et al. 2004
Laboratory scale (4 L), *S. cerevisiae*	52.44	–0.0762	0.514	0.702	Galaction et al. 2004
Laboratory scale (4 L), *P. chrysogenum* (pellet form)	0.193	0.0288	0.257	0.269	Galaction et al. 2004
Laboratory scale (4 L), *P. chrysogenum* (mycelial form)	33.59	–0.0463	0.94	1.012	Galaction et al. 2004

Table 5.3. *k_La prediction correlations as a function of bioreactor configuration and the type of microorganism studied*

5.3.3.2. *Extrapolation strategy based on a simplification hypothesis that is valid at small scales*

The hypothesis of O_2^* conservation between the two scales can be applied to small working scales, where the hydrostatic pressure difference remains negligible. In this case, equation [5.20] shows that the *OTR* conservation criterion also implies k_La conservation.

A relevant criterion for setting the injected gas flow rate can be based on maintaining the flow rate at the reactor surface, in other words, Q_g/T^2. A constant gas velocity u_g is then assumed for extrapolation.

Once the gas flow rate is fixed, and considering u_g to be constant in equation [5.22], the transfer coefficient k_La only depends on the aerated power P_g/V.

To determine its scaling law, P_g is expressed as a function of the power dissipated under non-aerated conditions, whose scaling law is controlled solely on the basis of agitation. Various correlations can be used to relate P_g to P_0, depending on the gas flow rate applied, which is set in advance here. These include the Calderbank correlation based on the aeration number, as shown in equations [5.24]–[5.26] (Calderbank et al. 1958).

$$N_A = \frac{Q_g}{ND^3} \quad [5.24]$$

$$\frac{P_g}{P_0} = 1 - 12.6 \cdot N_A \; for \; N_A < 0.035 \quad [5.25]$$

$$\frac{P_g}{P_0} = 0.62 - 1.85 \cdot N_A \; for \; N_A > 0.035 \quad [5.26]$$

Thus, keeping P_g/V constant, we deduce the value of P_0/V of the new scale, from the previous relations. Ultimately, equation [5.5] links this power in the non-aerated condition to the agitation frequency N to be applied, following the change of scale.

NOTE.– The approaches presented in this section are based on the estimation of an injected gas flow rate from the conservation of gas surface velocity. Other criteria can be used, such as maintaining the volume of sparged gas to volume of liquid per minute *vvm*; in other words, Q_g proportional to T^3, which then leads to an overestimation of k_La, thereby ensuring sufficient transfer for the process.

5.3.3.3. *Extrapolation strategy based on a digital resolution that is applicable to all scales*

In cases where the change of scale does not enable us to maintain, at least qualitatively, average O_2^* values from one scale to another, digital resolution is necessary. This is particularly important at very large scales, where significant O_2^* gradients can be observed along the height of the tank. In this case, the OTR must take both k_La and O_2^* changes into account.

We define the total tank surface pressure P^{tot} as the sum of the atmospheric pressure P^{atm} and any pressure applied by a back-pressure valve or sterility filters ΔP, according to equation [5.27]. Similarly, the pressure at the bottom of the tank is defined as a first approximation by considering the potential energy according to equation [5.28], where g is the gravity constant and H is the height of the fluid in the reactor.

$$P^{top} = P^{atm} + \Delta P \quad [5.27]$$

$$P^{bottom} = P^{tot} + \rho \cdot g \cdot H \quad [5.28]$$

Applying the pressures P^{tot} and P^{bottom} in Henry's equation, as shown in equation [5.21], we can then deduce O_2^{*top} and $O_2^{*\,bottom}$, representing the saturation values of oxygen concentration at the surface and bottom of the tank, respectively. Lastly, we approximate the property at a given scale by the mean value $O_2^{*\,mean}$ according to equation [5.29].

$$O_2^{*\,mean} = \frac{O_2^{*top} + O_2^{*\,bottom}}{2} \quad [5.29]$$

When the scale is changed, the determination of the injected gas flow rate must ultimately respect the equality of equation [5.30] between scales 1 and 2, to which the extrapolation method is applied.

$$OTR = (k_L a)_1 \cdot ((O_2^{*\,mean})_1 - O_2) = (k_L a)_2 \cdot ((O_2^{*\,mean})_2 - O_2) \quad [5.30]$$

Once the gas flow rate has been set, the steps for determining the agitation frequency condition then remain identical to the approach presented for small scales, as shown in section 5.3.3.2.

5.3.4. *Limits and assessment of bioreactor extrapolation strategies*

In biological processes (production of biomass, spores, acids, etc.), the main factors affecting process performance are oxygen transfer, characterized by the transfer coefficient (k_La), shear forces, power per unit volume (P/V), the Reynolds number (Re), mixing time (tm), etc. However, these parameters do not necessarily evolve in the same way when scaling up, due to possible variations in biological, hydrodynamic and physicochemical factors, and a compromise has to be found.

In aerobic cultures, oxygen transfer (k_La) is often the limiting stage on an industrial scale. One strategy for enhancing transfer is to increase the interfacial surface area of bubbles using porous devices (Hensirisak et al. 2002). However, coalescence and rupture phenomena need to be well characterized and controlled (Hasan and Zakaria 2011). Another strategy is to adjust the gas surface velocity, in relation to the injected gas flow rate, and the volumetric power. Lastly, it is also possible to promote transfer by injecting oxygen-enriched gas, with the partial pressure difference acting as the driving force for mass transfer.

The extrapolation methods presented in this chapter can be used to approximate the conditions required to maintain cultivation parameters that are close to the pilot scale. Nevertheless, it is important to keep in mind that the complexity of the interactions and relationships between the physical parameters governing bioprocesses calls for verification, and possible adjustment, during experimental validation on an industrial scale. One strategy for avoiding excessive deviation involves scaling up to a reasonable geometric ratio ($F \leq 10$) or adding intermediate study scales.

5.4. Conclusion

Scaling up bioreactors can involve several criteria. However, the choice of these depends on the nature of the microorganism being cultured and, more particularly, on its oxygen requirements and resistance to shear forces. Specific knowledge of these aspects is therefore crucial for successful extrapolation. At this stage of knowledge in the literature, no model is available to achieve perfect extrapolation between different scales (Marques et al. 2010). The expert approach is therefore indispensable when searching for a compromise between the criteria deemed to be most relevant and critical among the needs and sensitivity to biological, hydrodynamics and physicochemical constraints. It is also essential to minimize gradients in the fermenter (temperature, substrates, dissolved oxygen, pH, etc.). These gradients can affect the metabolism of microorganisms and, consequently, process performance (Nienow 1998; Vrabel et al. 2000; Junker 2004; Schmidt 2005; Hewitt and Nienow 2007).

5.5. References

Calderbank, P.H. (1958). Physical rate processes in industrial fermentation. Part 1: The interfacial area in gas-liquid contacting with mechanical agitation. *Trans. Inst. Chem. Eng.*, 36, 443–463.

Galaction, A.I., Cascaval, D., Oniscu, C., Turnea M. (2004). Prediction of oxygen mass transfer coefficients in stirred bioreactors for bacteria, yeasts and fungus broths. *Biochemical Engineering Journal*, 20(1), 85–94.

Garcia-Ochoa, F. and E. Gomez (2009). Bioreactor scale-up and oxygen transfer rate in microbial processes: An overview. *Biotechnology Advances*, 27(2), 153–176.

Ghasem, D.N. (2007). Bioprocess scale-up. In *Biochemical Engineering and Biotechnology*. Elsevier, Amsterdam.

Grenville, R.K. and Nienow, A.W. (2004). *Handbook of Industrial Mixing*. John Wiley & Sons, New York.

Hasan, N. and Zakaria, Z. (2011). Computational approach for a pair of bubble coalescence process. *International Journal of Heat and Fluid Flow*, 32(3), 755–761.

Hensirisak, P., Parasukulsatid, P., Agblevor, F.A., Cundiff, J.S., Velander, W.H. (2002). Scale-up of microbubble dispersion generator for aerobic fermentation. *Applied Biochemistry and Biotechnology*, 101, 211–227.

Hewitt, C.J. and Nienow, A.W. (2007). The scale-up of microbial batch and fed-batch fermentation processes. In *Advances in Applied Microbiology: Volume 62*, Allen, S.S., Laskin, I., Geoffrey, M.G. (eds). Academic Press, Chicago.

Hourcade, J.C. (1985). Signification et impasses d'un concept banalisé : les économies d'échelle. *Communication*, 42, 103–119.

Junker, B.H. (2004). Scale-up methodologies for Escherichia coli and yeast fermentation processes. *Journal of Bioscience and Bioengineering*, 97(6), 347–364.

Lidén, G. (2002). Understanding the bioreactor. *Bioprocess and Biosystems Engineering*, 24(5), 273–279.

Marques, M.P.C., Cabral, J.M.S., Fernandes, P. (2010). Bioprocess scale-up: Quest for the parameters to be used as criterion to move from microreactors to lab-scale. *Journal of Chemical Technology & Biotechnology*, 85(9), 1184–1198.

Möckel, H.O., Wolleschensky, E., Drewas, S., Rahner, H.J. (1990). Modelling of the calculation of the power input for aerated single and multistage impellers with special respect to scale-up. *Acta Biotechnologica*, 10(3), 215–224.

Neubauer, P., Cruz, N., Glauche, F., Junne, S., Knepper, A., Raven, M. (2013). Consistent development of bioprocesses from microliter cultures to the industrial scale. *Engineering in Life Sciences*, 13(3), 224–238.

Nielsen, J., Villadsen, J., Liden, G. (2003). Scale-up of bioprocesses. In *Bioreaction Engineering Principles*. Springer, New York.

Nienow, A.W. (1998). Hydrodynamics of stirred bioreactors. *Applied Mechanics Reviews*, 51(1), 3–32.

Nienow, A.W. (2009) Scale-up considerations based on studies at the bench scale in stirred bioreactors. *Journal of Chemical Engineering of Japan*, 42(11), 789–796.

Rodgers, T.L., Gangolf, L., Vannier, C., Parriaud, M., Cooke, M. (2011). Mixing times for process vessels with aspect ratios greater than one. *Chemical Engineering Science*, 66(13), 2935–2944.

Roustan, M., Pharamond, J.C., Line, A. (1999). Agitation. Mélange - Concepts théoriques de base. *Techniques de l'Ingénieur*, J3800.

Schmidt, F.R. (2005). Optimization and scale up of industrial fermentation processes. *Appl. Microbiol. Biotechnol.*, 68, 425–435.

Shukla, V.B., Parasu Veera, U., Kalkami, P.R., Pandit, A.B. (2001). Scale-up of biotransformation process in stirred tank reactor using dual impeller bioreactor. *Biochemical Engineering Journal*, 8(1), 19–29.

Smith, J.J., Lilly, M.D., Fox, R.I. (1990). The effect of agitation on the morphology and penicillin production of *Penicillium chrysogenum*. *Biotechnology and Bioengineering*, 35(10), 1011–1023.

Vrabel, P., van der Lans, R.G.J.M., Lyben, K.Ch.A.M, Boon, L., Ninenow, A.W. (2000). Mixing in large-scale vessels stirred with multiple radial or radial and axial up-pumping impellers: Modelling and measurements. *Chemical Engineering Science*, 55(23), 5881–5896.

PART 2

Examples of Bioreactor Applications in the Food Industry

Introduction to Part 2

Mohamed GHOUL
Laboratoire Réaction et Génie des Procédés (LRGP),
CNRS, Université de Lorraine, Nancy, France

The first part of this book explained the fundamental engineering principles required for the implementation and optimization of bioreactors used in the food industry. This second part is devoted to showing examples of these processes being applied, including the production of compounds and metabolites of interest used in the development of food formulations, such as proteins, colorants and organic acids or beverages such as beer.

This part is structured as three separate chapters, each examining a specific area in detail:

– *Beer production by yeasts in aerobic fermentation*: Chapter 6 presents the process of producing beer from malt by aerobic fermentation in the presence of specific yeast strains. The key factors affecting the performance of this complex process are presented and analyzed.

– *Biomass and dye production by microalgae culture in photobioreactors*: Chapter 7 looks at the cultivation of microalgae in photobioreactors as a promising source of proteins, natural pigments and numerous bioactive compounds. The operating conditions and type of microalgae used in each area are discussed.

– *Production of glutamic acid and ethanol using bacteria and yeast*: in Chapter 8, we detail the synthesis of glutamic acid production processes. The biochemical mechanisms involved in these productions, as well as the key fermentation parameters, are thoroughly described and discussed. This chapter also describes the economic optimization of bioreactors and examines the environmental impact analysis of these processes.

These examples enable readers to appreciate the multidisciplinary nature of these processes and the considerable potential offered by the application of bioreactors in the food industry.

6

Production of Fermented Beverages: Beer

Franck JOLIBERT[1] and Mohamed GHOUL[2]
[1]*Musée Français de la Brasserie, Saint-Nicolas-de-Port, France*
[2]*Laboratoire Réaction et Génie des Procédés (LRGP), CNRS, Université de Lorraine, Nancy, France*

6.1. Introduction

Beer is an age-old alcoholic beverage, and according to the World Health Organization (WHO), it is the most widely consumed in the world. It was originally produced in Mesopotamia and Egypt from sprouted cereals and aromatic plants, and fermented spontaneously. The Bavarian Purity Law of 1516 specified that beer was made from four ingredients: water, barley, hops and yeast.

Beer production is an anaerobic, exothermic process involving *Saccharomyces cerevisiae* yeast. In addition to transforming sugars into alcohols, the yeast contributes to the characteristic flavors and aromas of each type of beer.

The process of converting sugars into alcohol can be synthesized by the following equation:

$$C_6H_{12}O_6 + 2PO_4^{3-} + 2ADP \rightarrow 2C_2H_5OH + 2CO_2 + 2ATP \qquad [6.1]$$

Bioreactor Implementation in the Agro-Food Industries,

This reaction takes place in the absence of oxygen inside the yeast and is known as the Embden–Myerhof–Parnas or glycolytic pathway.

The conversion efficiency of sugars into ethanol is in the range of 50–60% by mass.

There are two types of fermentation: top fermentation and bottom fermentation. Top fermentation was practiced before industrial refrigeration became widespread. It takes place at a temperature of between 18 and 25°C for three days and produces fruitier beers. At the end of fermentation, the yeast rises to the surface. The beer it produces is known as "ale". There is no, or only a short, maturation stage. The yeast used does not ferment melibiose.

Bottom fermentation takes place at a temperature of between 10 and 18°C, and it lasts from five to 10 days. At the end of this fermentation, the yeasts flocculate and settle at the bottom of the fermenter. The main fermentation stage is followed by a maturation phase, and the resulting beer is called "lager".

Beer production became highly industrialized in the 1960s, with the emergence of major global brewing groups. Since 1990, however, this has not prevented the emergence of smaller breweries, known as "microbreweries" or "craft breweries". In France, for example, there are some 2,800 breweries, of which only 21 have an annual production capacity in excess of 100,000 hL of beer. The emergence of small-scale facilities does not just involve beer, but also spirits such as whisky, which is a hop-free, distilled beer.

This trend responds to consumer demand for beverages with a strong flavor, using locally produced raw materials (cereals, hops and other typical ingredients), including local microflora used in pure or mixed cultures. The use of these new *Saccharomyces* or non-*Saccharomyces* strains contributes to the creation of local beers.

6.2. Introduction to the brewing process

The brewing process is made up of several successive unitary operations, each with a precise role to play. The organoleptic and sensory qualities, alcohol yield and production cost of beer are closely linked to the conditions in which these operations are carried out. Figure 1.1 summarizes these operations.

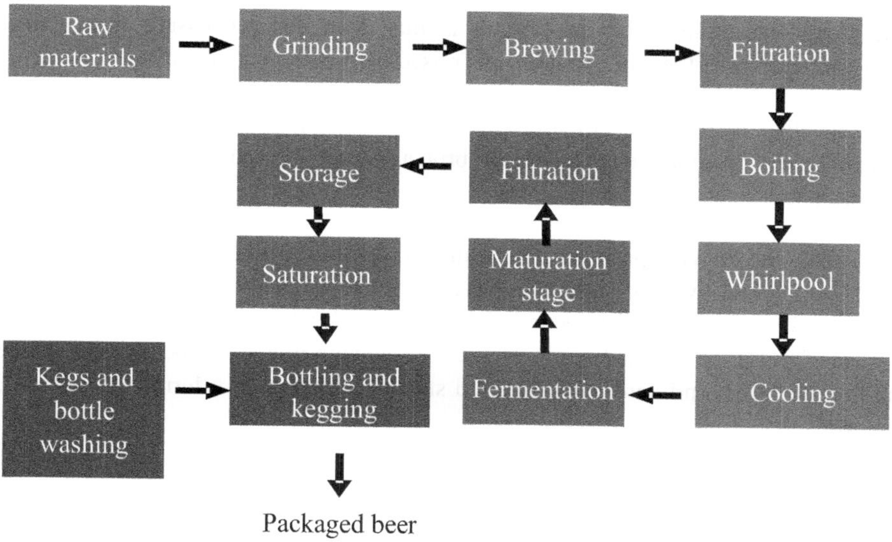

Figure 6.1. *Description of the different stages in beer production. For a color version of this figure, see www.iste.co.uk/ghoul/bioreactor.zip*

6.2.1. *Grinding and brewing operations*

Wort production begins with the grinding stage, using mills equipped with several pairs of rollers to grind the malt into floutabr, groats and husks. The mill is adapted to the filtration method that is subsequently used, as well as to the percentage of raw grains added. The resulting mill is put in hot water, in a vat known as a mash tun, to produce "mash".

Brewing produces a fermentable sweet juice called "wort" from friable malt. The brewer's role is to hydrolyze the starch and proteins using enzymes formed during malting. The efficiency of this step essentially depends on the temperature profile applied, which is 52°C for proteases and 64–72°C for β and α amylases. It also depends on the acidity and mineral content (especially calcium) of the water used. Brewing methods are specific to each brewery and can be classified into two types: decoction and infusion.

In the decoction brewing method, a portion of the mash is taken from the mash tun and boiled, after saccharification in a boiler attached to the mash tun, known as the mash tun boiler. This method is mainly used for processing raw grains. The temperature of the entire mash tun is raised by performing one or more mashes.

The other method, that of infusion, is simpler. It consists of carrying out the temperature steps corresponding to the proteases and amylases that are directly in the mash tun.

During the brewing phase, starch from malt and raw grains (corn, rice, sorghum, etc.) is liquefied by α-amylase enzymes. Complete saccharification translates into the transformation of all the starch into glucose, fructose, maltose and dextrins. The resulting juice is called mash. This is a dispersion of solids (spent grains, made up of malt husks and hard, non-liquefied starchy particles) in a fermentable liquid, the wort.

6.2.2. *Filtration and boiling operations*

The filtration stage is carried out in a filter tank or with a filter press. This separates the wort from the spent grains. To recover the fermentable extract that is bound to the spent grains, successive washings are carried out with hot water (80°C). The wort is recovered by filtration and the wash water is then collected in a tank, where the wort is cooked and hopped.

Boiling lasts 60–90 min, with an hourly evaporation rate of 3–5%, and results in the sterilization of the wort, the development of color through oxidation reactions, the elimination of undesirable odorant compounds, the dissolution and transformation of noble hop compounds, and the precipitation of protein complexes that cause hot break, also known as hot trub.

6.2.3. *Whirlpool operations and cooling*

The hot turbidity is then removed by means of a naturally centrifuging whirlpool tank, or a buffer tank followed by a centrifuge. The sterile wort thus obtained is cooled to a temperature of 2°C below the fermentation temperature, which depends on the type of yeast used.

The average composition of the wort in terms of the various major constituents is given in Table 6.1. This composition varies according to the type of malt used, the operating conditions applied in the brewing stage and the activity of the enzymes formed during barley germination.

		Weight %
Fermentable sugars	Fructose	1.2–3.3
	Glucose	9.3–10
	Sucrose	2.4–6.5
	Maltose	41–52
	Maltotriose	12.2–13
Non-fermentable sugars	Maltotetraose	2.1–5.9
	Maltopentaose	1.3
	Maltohexaose	1.9
	Dextrin (DP > 4)	11.5–23
	B-glucans, pentosans	0.1–0.3
Free amino nitrogen (100% barley malt)	180–230 mg/L	
Free amino nitrogen (30% raw grain)	140–155 mg/L	
Amino acids	145–150 mg/L	
Lipids	10–30 mg/L	
Polyphenols	150–300 mg/L	

Table 6.1. *Average major element composition of malt*

6.2.4. *Fermentation, maturation and saturation operations*

Beer production begins with a yeast propagation phase, also known as pre-culture or starter culture. This is carried out in rich media to ensure rapid growth and active yeasts. This phase is very important, as it not only affects the quality of the beer but also the duration of fermentation.

Main fermentation lasts 5–7 days at 18–25°C with *S. cerevisiae* for ale production, and at 10–18°C with *Saccharomyces pastorianus,* for lager production. Two phases are observed: the first corresponds to biomass growth, while the second, during which there is no growth, is essentially used for ethanol production.

Yeast activity converts around 75% of fermentable sugars into alcohol, carbon dioxide (CO_2) and heat.

Brewer's yeasts essentially belong to the *Saccharomyces* genus and are unicellular fungi, with a structure similar to that of higher order eukaryotic cells. They contain a cell wall, plasma membrane, mitochondria, Golgi apparatus, endoplasmic reticulum (ER), vacuoles and microbodies, secretory vesicles and a nucleus. The cell wall is mainly composed of carbohydrates surrounding the cell. It is a semi-rigid structure about 250 nm thick, representing 25% of the cell's dry weight.

Saccharomyces yeasts are generally ellipsoid in shape, but can also be spherical or cylindrical, with diameters ranging from 5 to 20 µm. *Saccharomyces* usually occur individually or in pairs but can sometimes occur in short chains or clusters. Almost all brewer's yeasts reproduce asexually by budding, creating bud scars composed of chitin when cells separate, but they can also reproduce sexually by conjugating cells of opposite mating types. In general, yeast clones itself no more than 20 times in its lifetime.

Compounds in beer	Organoleptic properties	Content in beer ($mg \cdot L^{-1}$)	Perception threshold ($mg \cdot L^{-1}$)
ALCOHOLS			
Ethanol	Strong alcohol	35,000–50,000	14,000
Active amyl alcohol (2 methyl butanol)	Iodoform alcohol	8–30	65
Isoamyl alcohol (3 methyl butanol)	Wine alcohol, banana, sweet	30–70	70
2 Phenyl-ethanol	Rose, bitter, perfume	10–60	125
ESTERS			
Ethyl acetate	Fruity, solvent, sweet	8–42	30
Isobutyl acetate	Banana, sweet, fruity	0.01–0.2	1.6
Isoamyl acetate	Banana, ester, solvent, apple	0.6–4.0	2.0
Ethyl caproate	Apple, aniseed, fruity	0.1–0.5	0.2
Ethyl caprylate	Apple, sweet, fruity	0.1–1.5	0.9
Phenylethylacetate	Rose, apple	0.1–0.8	3.8
ACIDS			
Butyric acid	Butter, cheese	0.6–3	2.5
Isovaleric acid	Cheese, old hops	1–3	1.5
Caproic acid (C6)	Fat, oil	1–6	8
Caprylic acid (C8)	Grease, oil	5–15	13–15
Capric acid (C10)	Rancid, soap	0.4–3	10
ALDEHYDES			
Acetaldehyde	Green apple	2–20	25
2-tr.-Nonenal	Paper, cardboard	1×10^{-5}–2×10^{-3}	11×10^{-5}
DICETONES			
Diacetyl	Rancid, butter	0.01–0.4	0.07–0.15
2-3 Pentadione		0.01–0.15	0.9
SULFUR COMPOUNDS			
Hydrogen sulfide	Rotten egg	1×10^{-3}–2×10^{-2}	5×10^{-3}
Dimethylsulfide (DMS)	Cooked vegetable	1×10^{-2}–0.2	33×10^{-3}

Table 6.2. *Perception thresholds for different beer compounds*

Secondary fermentation, or aging, allows the beer to mature, refine and reach carbon dioxide saturation. The process is carried out in a cold (0–3°C) environment for lagers.

Fermentation is a key stage in the beer production process. During this stage, the various constituents of the wort (sugar, nitrogen, lipids, etc.) are broken down through substrate-specific reactions. The average composition of a beer at the end of this fermentation stage is given in Table 6.2.

6.3. Substrate degradation pathways

6.3.1. *Carbon metabolism*

The main carbon sources used by yeast during fermentation are sugars, amino acids, lipids and alkanes.

6.3.1.1. *Sugar metabolism*

Sugars in the wort come in the form of monomers, oligosaccharides or polymers. Yeast makes very little use of polysaccharides. However, certain strains of *S. cerevisiae* are capable of utilizing pectin and have a complete pectinolytic complex. These yeasts are mainly used in oenology and cider production (Manshin 2022).

Sugars enter the cell by facilitated diffusion or active transport with group translocation. When transfer takes place by facilitated diffusion, sugars are found free in the cell cytosol. On the other hand, in the presence of active transport with translocation, phosphate sugars are found inside the cell (such as glucose-6-P and fructose-6-P).

All strains used in beer production processes are capable of utilizing glucose and fructose. Some species assimilate maltose and rarely maltotriose.

The use of sucrose occurs after its extracellular degradation by a parietal invertase. This invertase is also capable of hydrolyzing raffinose. Due to specific permeases, yeast strains are able to metabolize isomaltose and trehalose.

The use of maltose is regulated by the MAL gene system, which comprises five genes, including a maltose permease, an alpha-glucosidase (maltase) and three regulatory genes. Maltose induces the production of these two enzymes, which are also subject to dual control: they are retro-inhibited by glucose. Glucose causes catabolic repression, blocking their biosynthesis. These enzymes can also metabolize maltotriose, but at a much slower rate than maltose. Higher dextrins (DP4 and +) are not used by yeast.

Trehalose (glucose-alpha1-1glucose) is metabolized but can also be synthesized by yeast. From a metabolic point of view, trehalose acts as an anti-stress agent when present inside the cell, and yeast synthesizes it in response to physicochemical stress, such as alcoholic and thermal stress.

6.3.1.2. *Amino acid metabolism*

Amino acids play a dual role as sources of nitrogen and carbon. To be assimilated, they must first be incorporated into the cell. They enter by active transport through permeases, of which there are several types in yeast. All can be incorporated as they are for yeast protein biosynthesis.

However, it is essential for amino acids to be in free form if they are to be assimilated efficiently. Only small peptides are likely to be incorporated.

6.3.1.3. *Lipid metabolism*

Yeast normally synthesizes its own lipids. However, if it finds them in the culture medium, it incorporates them, without going through the synthesis stage.

6.3.2. *Nitrogen metabolism*

Nitrogen is mainly used to synthesize amino acids and the nitrogenous bases that make up nucleic acids. Proteins represent 47% of the dry weight of yeast. If free amino acids are present in the medium, yeast will preferentially use them; it will therefore not use sugar for the carbon chain of its proteins.

Group	Rate of amino acids absorption	Amino acid symbol
1	Immediate absorption and total elimination	Glu, Asp, Asn, Gln, Ser, Thr, Lys, Arg
2	Absorbed gradually during fermentation	Val, Met, Leu, Ileu, His
3	There is a considerable time lag in the absorption of these amino acids, and their assimilation coincides with the disappearance of group A amino acids.	Gly, Phe, Tyr, Trp, Ala, NH_3
4	Proline is poorly absorbed	Pro

Table 6.3. *Classification of amino acids according to their wort absorption rate under brewing conditions*

Amino acids are classified into four groups according to their rate of assimilation (Table 6.3).

– *Organic nitrogen*: organic nitrogen mainly comes from protein nitrogen, either in the form of free amino acids, or in the form of oligopeptides, but rarely in the form of proteins.

– *Amino acids/peptides*: small peptides (two to four amino acids) are relatively well used, while those with more than seven amino acids are not.

If free amino acids are present in the medium, they will be preferentially used for protein synthesis. However, only those from Group 1 are metabolized efficiently; in other words, they are used to produce other amino acids or other metabolic compounds.

Proteases are sometimes added. This increases alcohol yield, as the glucose not used for protein synthesis is used in the fermentative pathway. In contrast, this addition affects the final protein content, foam quality and nutritional quality of the beer.

6.3.3. *Phosphorus metabolism*

Phosphorus is the third most important element after carbon and nitrogen. Yeast is able to assimilate inorganic phosphate through an active transport mechanism. It is therefore autotrophic for this substrate. The PO_4^{2-} ion is the most easily assimilated source of phosphate. Phosphorus enables yeast to produce cellular components such as nucleic acids.

6.3.4. *Sulfur metabolism*

Sulfur is the fourth most important element after carbon, nitrogen and phosphorus. Yeast is capable of assimilating inorganic sulfur and is therefore autotrophic for this element. As inorganic sulfur, only the SO_4^{2-} form is used by yeast; it does not use the elemental sulfur (S) or S^{2-} forms. The form most commonly used as a sulfur source is $SO_4^{2-}(NH_4^+)^2$, which simultaneously provides sulfate and ammonium. As far as organic sulfur is concerned, yeast only uses the sulfur-containing amino acids, cysteine (Group 3) and methionine (Group 2). Methionine is more readily assimilated than cysteine.

6.4. Metabolite synthesis pathways

The main products of alcoholic fermentation are, of course, ethanol and carbon dioxide. Together, they account for 90–95% of the sugars used during growth for biomass production, as well as for the formation of by-products that are essential for beer's aroma.

The major by-products of alcoholic fermentation are glycerol, succinate and acetate, which use up around 4–5% of the remaining carbon substrate. In addition, acetaldehyde, fatty acids such as hexanoic acid, octanoic acid, decanoic acid, etc., as well as alcohols containing more than two carbons, are produced. These constituents are the source of beer's flavor.

6.4.1. *Synthesis of higher alcohols*

During the anaerobic fermentation stage, in addition to ethanol, higher alcohols with long carbon chains are synthesized. Among these alcohols, isoamyl alcohol (2-methyl-1-butanol), the precursor of the banana aroma and phenylethyl alcohol or tyrosol, the precursor of the rose aroma, are often present in doses above the perception threshold.

Other higher alcohols, such as propanol, butanol and isobutanol, are generally present in quantities below the perception threshold. Two metabolic pathways contribute to the formation of these higher alcohols: the catabolic pathway (known as the Ehrlich reaction) and the anabolic pathway.

The contribution of the catabolic pathway depends on the quantity and type of amino acids present. The anabolic pathway is favored when the medium contains little nitrogen, or when the main source of nitrogen is ammonium or urea.

The formation of higher alcohols is favored by conditions that lead to high yeast multiplication, such as high fermentation temperature, high amino acid levels and high wort aeration. However, it is reduced in the presence of low temperatures, increased yeast doses, high pressure during fermentation and low wort aeration.

Increasing the concentration of amino acids improves the efficiency of the Ehrlich pathway. By combining endo and exo-proteases during brewing to selectively release amino acids, aromas are enhanced. Under these conditions, wort fermentation produces around 10% more isoamyl alcohol, 17% more isoamyl alcohol and leucine-derived isoamyl acetate, when compared with an indicator (Lin et al. 2022).

6.4.2. *Aldehyde synthesis*

These compounds play an important role as precursors to alcohols, and as residual products in beer. Their presence in beer varies according to the operating conditions of fermentation, resulting in variable concentrations. For example, acetaldehyde content can increase by up to 20 ppm as temperature rises.

Trans-2-nonenal, which has aromas of paper and cardboard, is representative of the long-chain aldehydes that are characteristic of fatty acid oxidation.

6.4.3. *Synthesis of organic acids*

Fatty acids such as caproic, capric and caprylic acids are the main metabolites synthesized by yeast during fermentation. The main fatty acids found in beer are derived from yeast metabolism. Unlike those present in wort cloudiness, they are saturated and have medium-length chains (C6-C8-C10). Fatty acids are characteristic of yeast strains, particularly in the autolysis phase. Other organic acids, such as acetic acid and lactic acid, mostly originate from bacteria. These molecules influence the beer's acidity.

6.4.4. *Ester synthesis*

Esters are synthesized during fermentation and are an important group of volatile compounds that influence beer flavor. Ester synthesis is linked to lipid metabolism in yeast. The most significant esters are ethyl acetate (solvent), isoamyl acetate (banana aroma) and ethyl caproate (apple aroma).

Ester formation is favored by high density, low wort aeration, reduced fermentation temperature, and low fatty acid levels in the must. Esters are by-products of lipid synthesis and are formed by the enzymatic esterification of fatty acids and higher alcohols from acetyl CoA. These esterification reactions between alcohols and acids create a group of volatile compounds that influence beer flavor.

Ester production is part of a yeast's "signature", but it is also closely linked to the operating conditions applied during the fermentation stage, such as wort density and aeration (Paszkot et al. 2023).

Ethyl esters are esters formed by condensation between a fatty acid and an alcohol. The main ones are ethyl butanoate (apple and strawberry flavors), ethyl hexanoate (green apple and banana aromas), ethyl octanoate (banana, pineapple and pear aromas), ethyl decanoate (floral, grape aromas) and ethyl dodecanoate (floral aroma). Their concentrations in beer are generally close to (or above) perception thresholds. These molecules are formed during fermentation and require alternating aerobic and anaerobic steps for their synthesis.

6.4.5. *Diketone synthesis*

Diacetyl is the main compound in this class, with a perception threshold of 0.15 mg/L (cream and butter aroma), followed by 2,3-pentanedione, which is 10 times less fragrant. They are derived from yeast metabolism during amino acid synthesis.

The formation and reduction of diketones takes place in three stages. The first is the formation of keto acids (2-acetolactate and 2-acetohydroxybutyrate) from pyruvic acid. These reactions are catalyzed by yeast during amino acid synthesis, and are influenced by yeast strain, seeding rate and wort aeration. The second stage leads to the formation of diketones, following the oxidative decarboxylation of keto acids, which occurs independently of the yeast. This stage is influenced by low pH (4.2–4.4), high temperature and aeration. The final stage is a reduction of diketones according to the following mechanism:

$$\text{Diacetyl} \rightarrow \text{acetone} \rightarrow \text{butanediol} \qquad [6.2]$$

In top fermentation, yeast is able to reduce diacetyl 10 times faster than it is formed. This diacetyl reduction is favored by high temperature and high yeast concentration.

6.4.6. *Synthesis of sulfur compounds*

Sulfur compounds come from raw materials such as water and amino acids, but also from yeast metabolism during fermentation. The two main sulfur compounds are dimethyl sulfide (CH_3-S-CH_3) and hydrogen sulfide (H_2S).

Dimethyl sulfide and its precursor (S-methyl methionine) originate from raw materials, notably sulfate in water and sulfur-containing amino acids in cereals, as well as sulfur from yeast metabolism and bacterial infections in wort.

All brewer's yeasts produce a certain amount of H_2S during fermentation, as part of normal amino acid metabolism. In the case of top fermentations, H_2S is effectively eliminated by leaching, from the CO_2 produced during the active fermentation stage, as well as by reabsorption by the yeast after complete attenuation.

However, in the case of bottom fermentations, which are slower and take place at lower temperatures, CO_2 leaching is less significant. Similarly, lager yeast, which is highly flocculent, does not reabsorb H_2S as effectively. Small amounts of H_2S, in excess of detection limits, produced by these traditional lager yeast strains, are considered to be a significant part of the normal sensory profile of many lager beers. However, poor wort quality or brewing techniques can lead to high levels of H_2S and an undesirable rotten-egg aroma. This can be avoided by adding nutrients to the wort, or by giving the beer sufficient time for the yeast to reabsorb H_2S before transfer or filtration.

The more delayed the sulfur dioxide is produced, the lower the final SO_2 content. If yeast growth is slow, sulfur dioxide production will also be delayed, resulting in a low SO_2 content (10 mg/L).

6.5. Effects of physicochemical factors on the beer production process

Many factors influence the fermentation stage of beer: the composition and concentration of substrates in the wort, the yeast strain and its physiological state, the wort aeration, the quantity and method of adding yeast to the wort, the fermentation temperature, the quality of wort/yeast homogenization, the geometry and size of the fermenter, hydrostatic pressure, circulation speeds and additives/ manufacturing aids.

These additives can be composed of long-chain polysaccharide cellulose, ammonium diphosphate, ammonium sulfate, vitamin B1 and hydrated silica gel. Polysaccharide cellulose has a special structure. This structure enables the yeast to remain homogeneously suspended in the fermenters as long as fermentable sugars are available.

6.5.1. *Effect of density*

The success of fermentation is assessed by measuring the density, which is an indication of the proportion of sugars consumed during fermentation between the initial density (DI) and the final density (DF). A yeast with a seemingly high attenuation value produces a beer with a drier, less malty and potentially higher alcohol content. Conversely, if attenuation is low, yeast will consume less or no complex sugars such as maltose and maltotriose, resulting in a fuller-bodied, more malty and potentially less alcoholic beer.

Attenuation is calculated according to the following equation:

$$\text{Attenuation} = \left(\frac{\text{DI}-\text{DF}}{\text{DI}-1}\right) \times 100 \qquad [6.3]$$

During fermentation, the reduction in density results from the disappearance of fermentable sugars and the appearance of alcohol. This is due to the fact that sugar, whose density is greater than that of water, is transformed into less dense alcohol and CO_2 gas.

The final density of a beer is linked to the quantity of fermentable, but unfermented, sugars in relation to non-fermentable sugars (dextrins). This ratio will depend on the efficiency of starch hydrolysis during the brewing stage, the quality of the grinding, the pH during brewing (between 5.6 and 5.8), the quality of the malt, the presence of calcium ions in the water and keeping to the temperature steps during brewing.

It is well known that the fermentation process not only generates alcohol and CO_2, but also a large number of compounds that have a strong influence on the productivity and quality of the beverage. To guarantee good beer quality, the brewer must therefore minimize the metabolites that have a negative impact as much as possible, while optimizing the quantity of those with a positive effect.

In order to increase productivity and reduce beer production costs, high-density fermentation has been proposed as an alternative solution to conventional fermentation. High-density fermentations offer a number of advantages, including increased production capacity without modifying the production structure, the possibility of producing a wide range of beers after dilution with de-aerated water and the addition of additional aromas, reduced energy costs per liter of beer for the fermentation stage, limitation of contamination risks thanks to high osmotic pressures and high ethanol concentrations, as well as reduced water consumption and the amount of effluent to be treated.

In order to achieve high-density fermentations, it is recommended to use a high content of malts that are rich in amylolytic enzymes, mashing with a water/malt ratio of 3:1, the use of special malts (notably caramel), a shortened proteolytic plateau and long saccharification to obtain maximum fermentable sugars, followed by prolonged boiling to improve protein flocculation. Biological wort acidification is also recommended; in other words, the addition of mash that is spontaneously acidified by lactic acid bacteria or phosphoric acid (Millet 2019).

A 10–20% increase in degrees Plato (°P) increases ethanol production by approximately 1.9 times, acetates and esters by three to four times, and higher alcohols by 1.5–2 times. Acetaldehyde decreases by 0.6. The ester/ethanol ratio is multiplied by 2, which, after dilution of the resulting beer, makes it possible to adjust the alcohol content while maintaining a high ester content.

To optimize the organoleptic profile, the same authors recommend raising the fermentation temperature from 12 to 18°C in bottom fermentation and from 18 to 24°C in top fermentation, in order to increase the content of higher alcohols, ethyl acetates and isoamyl acetates (mainly in bottom fermentation), and to slightly aerate the wort.

6.5.2. *Effect of temperature*

Temperature influences growth and metabolite production rates. The higher the temperature, the faster the growth. The choice of temperature is therefore crucial. The optimum growth temperature for *S. cerevisiae* is between 12 and 33°C. However, this depends on strain, medium and culture conditions.

As the temperature rises, the yeast is also subjected to greater stress. Ultimately, above 25°C, the yeast becomes significantly more sensitive to osmotic pressure, leading to an increase in residual sugar concentration under these conditions.

6.5.3. *Effect of acidity*

The increase in acidity causes a decrease in the growth rate, until it is canceled out at a concentration equal to or greater than 5 g/L, expressed in sulfuric acid equivalent. Titratable acidity is therefore measured in parallel with pH monitoring. For some strains, even at pH levels that are compatible with yeast growth, acidity can inhibit the production of key metabolites for beer aroma.

Despite its inhibiting effect, acidity is sometimes useful in limiting the development of bacterial contaminants. An acidity of between 1.5 and 2.5 g/L would be a compromise between the bacteriostatic effect and optimal yeast development and beer quality.

6.5.4. *Effect of inoculation rate*

The inoculum of active yeast added to the wort to initiate fermentation is a key factor. Successful fermentation depends on the quality and quantity of the inoculum, which should generally be around 10^6 cells/mL. This seeding rate favors the consumption of maltose and other sugars that are available in the wort. The quantity and rate of inoculum are also very important factors in controlling the duration of fermentation.

To reduce fermentation time, we recommend transferring the yeast from the propagation medium to the fermentation medium, when the yeast is in its exponential growth phase. During this phase, yeast metabolism is at its peak.

To improve the inoculum, a culture is seeded with yeast from the previous fermentation. Yeasts adapt to culture conditions over the generations. In this way, cell membranes become better adapted to the environment, thanks to changes in lipid compounds and the quantity of sterols present.

To improve the fermentation stage, beer is sometimes added to the wort on the second day of fermentation (10% by volume) from a beer in storage in order to kick-start a slow secondary fermentation. This technique overcomes the problems of yeast dormancy during the maturation phase, and also helps to reduce diacetyl and acetaldehyde levels. This method is mainly used in German breweries and keeps the beer in a reducing environment. It is called “Krausen”. The use of yeast from a previous fermentation is limited to a maximum of seven generations.

6.5.5. *Effect of pressure*

Pressure affects exchanges between the medium and the cell. A pressure setpoint is often imposed during fermentation to offset the negative effect of rising temperature. This pressure setpoint is generally between 0.5 and 3 bars, with a typical value of 2 bars. Pressurization promotes CO_2 dissolution and better saturation of the beer. This reduces the need for CO_2 during its carbonation.

6.5.6. *Effect of agitation and fermenter configuration*

Beer tanks are agitated without the need for a mechanical agitator, thanks to the release of carbon dioxide. Agitation plays a key role in improving the transfer of matter and heat between phases. It also helps keep yeast in suspension and promotes sedimentation of flocculated yeast at the end of fermentation, without off-gassing.

The shape of the equipment, and more specifically the height/volume ratio, also affects fermentation. A fermenter characterized by a high height and a small diameter is subject to more intense agitation, accompanied by greater gas development and high pressure at the bottom of the fermenter. These conditions favor the production of higher alcohols, organic acids and cell multiplication. On the other hand, they reduce ester synthesis.

The roughness of the fermenter also plays a role. It promotes turbulence during cleaning and the attachment of yeast to the wall. The metabolism of fixed yeast is enhanced compared to suspended yeast.

The filling time of the fermenter with several successive brews should not exceed 48 h; it should ideally be 24 h. Similarly, during the latter period, the tendency is to carry out fermentation and aging in the same fermenter with a usable volume of 80% to take account of foaming.

To preserve the aromatic value of the esters, it is preferable to start fermentation in a horizontal cylindrical tank until 7% °P is obtained, and then finish fermentation in a cylindrical–conical tank.

The shape of the cylindrical–conical tanks has evolved to obtain a domed top, which is better suited to high pressures. The bottom is conical with a 60° angle to facilitate yeast settling.

Three levels can be distinguished in the bottom of the tank. The lower level contains dead or inactive yeast, mixed with cloudiness, and is disposed of as spent yeast. The middle level represents the harvest; in other words, healthy yeast that can be stored and reused. The top level contains the smaller, less active yeasts, also known as spent yeast.

To achieve reproducible beer quality, fermenter automation has become widespread. It enables real-time monitoring and control of temperature, pressure, density drop, volume of gas produced, pH evolution, yeast racking and cleaning-in-place.

6.5.7. *Inhibition phenomena*

– *Substrate inhibition*: high carbon substrate concentrations in excess of 150 g/L inhibit cell growth and fermentation capacity (Riess 2012). This phenomenon can be explained by inhibition of glycolysis enzymes or by high osmotic pressure, accompanied by lower water activity.

– *Inhibition by metabolites produced*: the growth-inhibiting effect of the main metabolites synthesized (acetic acid, formic acid, lactic acid, acetaldehyde, glycerol, etc.) was analyzed by Maiorella et al. (1983). Inhibitory concentrations were 450, 38, 7.5, 5 and 2.7 mg/L for glycerol, lactic acid, acetic acid, acetaldehyde and formic acid, respectively. Hexanoic, octanoic and decanoic fatty acids are the main inhibitors under batch fermentation conditions. This inhibition is estimated at around 1% per mg/L of fatty acid, with decanoic acid being more toxic than octanoic acid.

– *Ethanol inhibition and the notion of alcohol tolerance*: ethanol, produced during fermentation, is toxic to yeast and can lead to prolonged fermentations or even stuck fermentations. There are various effects of ethanol on *S. cerevisiae*: reduced growth rate, reduced viability, reduced fermentation capacity, destabilization of the plasma membrane and increased membrane permeability, stimulation of membrane ATPases, resulting in a loss of energy available to the cell, and inhibition of glucose and nitrogen transport. It has been shown that even at low ethanol concentrations, between 4 and 6% (v/v), yeast becomes more stressed. The inhibitory effects become more pronounced when ethanol production exceeds 10% (v/v).

6.5.8. *Infection by other microorganisms*

Very few microorganisms are able to grow in beer. This is due to the low pH (3.9–4.4), the presence of ethyl alcohol and hop resins, as well as the lack of oxygen. Infections are therefore essentially limited to wild yeasts or Gram-positive bacteria.

Bacterial infection can originate from air, water or inoculum. However, pathogens cannot survive in beer due to the low pH (3.9–4.4) and the unfavorable physicochemical environment. A low dose of bacteria in beer is not always detrimental to taste, but in the event of excessive proliferation, it can affect the beer's taste quality and render it totally undrinkable.

Two categories of bacteria are encountered: lactic acid bacteria and acetic acid bacteria.

The lactic acid bacteria found in beer are mainly *Pediococcus* and *Lactobacillus*, which produce lactic acid by fermenting simple sugars under anaerobic conditions. These bacteria are also tolerant of hops and ethanol. They can cause acidification, sourness, cloudiness and off-flavors. They mainly come from poorly cleaned equipment or the fermenter environment. Infection by lactic acid bacteria is a relatively slow process. These bacteria can degrade both fermentable and non-fermentable sugars. They are often responsible for exploding bottles that have been filled up.

The main bacteria responsible for releasing acetic acid are *Acetobacter* and *Acetomonas*, which are present in the air. These bacteria can operate under all pH conditions encountered during beer brewing, and they are not inhibited by hops or alcohol. *Acetobacter* oxidizes ethanol in the presence of oxygen, producing acetic acid (vinegar). These bacteria are powerful oxidizers, responsible for unpleasant tastes, aromas and cloudiness. Contamination is often visible as an oily-looking film. These types of bacteria generally enter the wort during transfer or aeration, as the presence of oxygen favors their growth and activity. To avoid infection, it is therefore necessary to carry out a fermentation that rapidly consumes almost all the oxygen available in the medium.

Symptoms of acetic bacteria contamination include a sour taste, cloudy beer, an abnormal drop in pH, and an unusual drop in alcohol content, as ethanol is oxidized to acetic acid.

Other bacteria can contaminate beer, notably the recently discovered Gram-negative *Pectinatus* and *Megasphaera*. Wild yeasts also contribute to the degradation of beer's organoleptic quality (Thompson and Dawson 2023).

Fungal contamination is often reported in malt. The most common are *Fusarium*, *Cladosporium*, *Alternaria*, *Vishniacozyma* and *Sporobolomyces*. They are responsible for the phenomenon of spontaneous foaming, known as "gushing".

6.6. New trends in the beer fermentation process

The fermentation of beer is historically linked to the use of *S. cerevisiae*. These yeasts are classified into different lineages according to their genomic structure. *Saccharomyces* cannot grow on lysine as the sole nitrogen source.

Two types of *Saccharomyces* are mainly used in brewing: *S. cerevisiae* and *Saccharomyces pastorianus. S. cerevisiae* yeast is used for ales, obtained by top fermentation at temperatures between 18 and 25°C. These yeasts assimilate maltose and maltotriose, and they tend to rise to the surface of the vat. *Saccharomyces pastorianus* yeast is used to make "lager" beers, which are obtained by bottom fermentation at temperatures between 10 and 18°C.

These two yeasts also differ in their respiratory power, their degree of sporulation, their ability to ferment glyceraldehyde and their capacity to produce H_2S. In addition, there are differences between these two yeasts in terms of the production of by-products in different quantities, such as esters and higher alcohols.

6.6.1. *Yeast strain improvement*

In recent years, a great deal of work has been carried out to improve brewing strains in order to obtain the desired organoleptic qualities, increase yields of aromatic compounds or develop yeasts resistant to variations in their environment. These improvements are based on interspecies hybridization. Recently, classical non-GMO (genetically modified organisms) hybridization methods have been used to select new hybrid lager strains that are distinct from traditional lines. These new strains are allotetraploid, with four sets of chromosomes: three from *S. cerevisiae* and one from *Saccharomyces eubayanus*. The parameters guiding the selection are attenuation, alcohol tolerance, flocculation and organoleptic quality (Carmen-Rodica 2022 et al; Kato and Takahashi 2022).

A first commercial example of these strains is a lager with a greater contribution from the *S. cerevisiae* sub-genome. It features higher temperature tolerance, more robust and faster fermentation, and a unique organoleptic profile. Indeed, it is characterized by its ability to produce low levels of diacetyl and H_2S by counter-selecting sulfur-specific metabolic genes, while maintaining the cryotolerance given by the *Saccharomyces eubayanus* subgenome. This yeast is capable of valine uptake and diacetyl levels similar to those of ale strains, contributing to shorter maturation times. Hybridization tests have also been carried out between *Saccharomyces jurei* and *S. cerevisiae* (Astola et al. 2023).

These improvements have led to the commercial availability of numerous strains for the production of typical beers (pilsners and special lagers, blond and brown ales, high-alcohol ales, Kolsch, Saison, white beers, India pale ales, porters, stouts, imperial, barley wines, sour beers, etc.). Strains with specific enzymes for the decarboxylation of phenolic acids, producing 4-vinyl guaiacol (4VG), which are responsible for spicy, clove-like aromas, have also been developed (de Ruijter et al. 2023).

Non-*Saccharomyces* yeasts such as *Brettanomyces*, *Pichia*, *Hanseniaspora*, *Metschnikowia*, *Naumovozyma*, *Debaryomyces*, *Hanseniaspora*, *Wickerhamomyces* and *Torulaspora* have been industrially tested for beer production. *Brettanomyces* spp. yeasts are still regarded as spoilage organisms responsible for producing off-flavors in fermentations, but they can contribute desirable organoleptic characteristics and interesting aromas to hopped beers. They have also been explored for their ability to give high ethanol yields and tolerance to low pH. *Brettanomyces* is thus marketed for secondary fermentation in bottles or kegs, producing phenols 4-ethylguaiacol and 4-ethylphenol (Bruner 2022; Iturritxa et al. 2023).

Like *Brettanomyces*, the *Hanseniaspora* genus can add desirable aromatic qualities in mixed-culture fermentations such as lambic beer, but it ferments slowly and is not ideal at low temperatures. *Pichia* spp. are also considered spoilage organisms in beer, as they contribute to off-gassing in the bottled product, but may be desirable in some mixed fermentations and in native beer. Beers produced with *Metschnikowia* spp. are promising for low-alcohol and non-alcohol products, but consumer preference calls for their use in mixed cultures (Nieto-Sarabia et al. 2023).

Numerous other strains (*Torulaspora* spp., Cyberlindnera saturnus, *Torulaspora delbrueckii, Linda Kveik*) have been identified for their ability to produce specific aromas. These yeasts are used in craft beer production processes, but rarely on an industrial scale (Kayadelen et al. 2023).

6.6.2. *Co-fermentations*

Another growing trend is the use of mixed cultures with heterofermentative lactic acid bacteria, for example, to create safe beers with full-bodied, well-balanced acidity. Co-fermentation is achieved by taking a portion of the wort before boiling, so as to ensure fermentation by the lactic acid bacteria, and adding it to the boiling tank along with the hops. The bacteriostatic properties of hops inhibit lactic acid bacteria. It transforms glucose into 90% lactic acid and acetic acid.

The use of *Torulaspora delbrueckii* in combination with an ale-type *S. cerevisiae* produces beers with ethyl alcohol contents ranging from 5.5% (v/v) to 6% (v/v). Among beer's aromatic compounds, the main volatile compounds are acetaldehyde, n-propanol, 3-methyl-butanol, 2-methyl-butanol, ethyl acetate, isoamyl acetate, 2,3-butanedione and 2,3-pentanedione. The total amount of higher alcohols is higher in the *S. cerevisiae* sample beer than in mixed fermentations.

Total ester levels are higher in the mixed culture beers than in the sample beer. Sensory evaluation shows a better sensory profile for mixed cultures (full-bodied, hoppy). *Torulaspora delbrueckii* strains marketed for winemaking can therefore be used for brewing (Kayadelen et al. 2023).

New developments include the use of *Lactobacillus* and *Pediococcus* to acidify beers. Lastly, certain hybrid strains, such as *Philly Sour*, produce both lactic acid and alcohol to avoid the risk of cross-contamination.

6.6.3. *Fermentation processes*

6.6.3.1. *Addition of hops in the secondary fermentation phase*

This trend has been observed recently, and involves the addition of hops, usually in the form of 10 g/L pellets (dry hopping), which do not add bitterness to the beer, but reinforce the aromas that are usually lost during boiling. These additions also give the beer fruity notes and increased astringency.

The best hops to use for dry hopping are noble hops with rich aromas, such as lemony hops ("cascade" and "citra"), citrus and exotic fruit hops ("simcoe", "centennial" and "mosaic") and resinous, spicy hops ("saaz" and "east kent golding").

Hops also provide sugars and an enzyme complex that contribute to the formation of complementary fermentable sugars. Recent research has shown the interactions of certain strains with specific hop compounds through the enzymes β-glucosidase and β-lyase. These interactions enhance the beer's flavors and aromas.

6.6.3.2. *Low-alcohol beers*

The market for low-alcohol beers (<2.8%vol), non-alcoholic beers (<1.2%vol) and alcohol-free beers (<0.5%vol) is strongly expanding. There are a number of ways to reduce the alcohol content of beer during the production process or by eliminating it afterward by, for example, distillation.

Methods based on limiting alcohol production during the fermentation stage all act on fermentation conditions to limit the transformation of sugars into alcohol. Various processes are available:

– *Sugar-free fermentation*: beer is brewed slowly at low temperature to avoid producing fermentable sugars. Less active yeasts, such as *Saccharomyces ludwigii*, can also be used (Jackowski et al. 2023).

– *Rapid fermentation*: fermentation is stopped abruptly with extreme cold after a conventional start. The disadvantage of this method is that it produces a low-sparkling beer that requires carbonation. Another possibility is to initially use less malt, thus reducing the amount of fermentable sugars, but supplementing with special malts to retain sufficient flavor. Another option is to use wort that is cooled to a temperature of less than 0.5°C to +1°C, with the addition of water-washed yeast that does not consume amino acids, but only degrades glucose and fructose.

– *Extrapolation of the Barrel and Victoria method (Australia)*: these Australian authors suggest brewing a first regular beer at 4.2% v/v and a second beer at 0.8%, both with the same hopping process, but with the second being pH adjusted. Combining the two results in a 2.5%vol beer.

– *Co-fermentation of Lactobacillus fermentati and Candida subsufciens with Lactobacillus plantarum*: this strategy is promising for beer production below 0.5% ethanol. Due to the limited fermentation capacity of the yeasts and, consequently, lower ethanol production, fermentation time is extended when compared to a sample yeast, such as *S. cerevisiae*. The beer shows a relatively high concentration of lactic acid, lower levels of residual sugars and a reduction in the amount of diacetyl below the sensory threshold. This process is currently being developed to obtain a low alcohol content with a good aromatic profile.

– *Process developed by EvodiaBio*: this company presents a patented fermentation method that produces the aromas that are usually missing when making non-alcoholic beers. The company uses yeast cells that secrete individual aroma components and combines them to mimic the desired aroma profiles of different hops.

6.6.3.3. *Refermentation in the bottle*

This is an age-old method that has been reintroduced thanks to the development of microbreweries. It involves triple fermentation, similar to the carbonation quality developed in Champagne. This process takes place in a bottle or barrel, with sugar and yeast added to the container. The effectiveness of refermentation depends on the physiological state of the yeast, which is in an inhibiting environment. As rapid bottle fermentation is desirable, the yeast must have a reserve of energy and a high level of enzymatic activity. This is why we use yeast in its exponential growth phase, that is to say, between the second and third day of fermentation.

To avoid autolysis tastes, the inoculation rate should be reduced. This compromise is found for a value of 10^5 to 5×10^5 cells/mL and an oxygen level of 2 ppm.

During the bottle fermentation stage, amino acid consumption is very low, so there is little production of higher alcohols. Similarly, the level of esters, which result from the esterification of an alcohol with an acid, is also very low.

6.6.4. *Development of rapid control techniques*

In addition to conventional techniques for monitoring the physicochemical parameters of fermentation, recent years have seen the introduction of rapid methods (<3 h) using PCR (polymerase chain reaction). These enable the detection of spoilage flora (hop-resistant *Pediococcus* and *Lactobacillus*, *Brettanomyces bruxellensis*, *Saccharomyces cerevisiae var diastaticus* responsible for gushing, *Megasphaera* and *Pectinatus*, etc.) in beer.

Analyses are carried out on all types of matrices, such as wort, beer, raw materials and inoculum.

Similarly, ATP-metry has been introduced to the field of beer production. This enables automatic colony counting and facilitates analysis of trends in the development of microorganisms that are present in the medium.

6.6.5. *Conclusions and perspectives*

Given the large number of developing craft breweries and a promising market, it seems that the future lies in the use of terroir yeasts, with a wide range of specific yeasts for the different types of beer sought after by new amateur consumers.

At the same time, yeasts selected by hybridization, which also develop specific aromatic capacities, are being increasingly used by major breweries.

In order to reduce energy and water consumption, as well as the environmental impact of these processes, rapid fermentation yeasts that tolerate high stress conditions (sugars, alcohol, temperature, acidity, etc.) are increasingly being researched and tested. Numerous strains from the wine sector are being assessed according to the specific sugar composition of wort, and the presence of alcohol and hops. The *Lachancea thermotolerans* species, for example, has shown great potential in terms of both aromatic profile and alcohol content.

6.7. References

Astola, A., Durán-Guerrero, E., Belén Díaz, A., Lasadnta, C., Castro, R. (2023). Impact of the genetic improvement of fermenting yeasts on the organoleptic properties of beer. *European Food Research and Technology*, 249, 1677–1687.

Bruner, J.R. (2003). Investigating yeast and hop interactions during the fermentation of beer. Thesis, University of California, Davis [Online]. Available at: https://escholarship.org/uc/item/7p40k3x5 [Accessed 12 January 2023].

Carmen-Rodica, V., Barbulescu, J.D., Matei, F., Begea, M., Tudor, V., Frincu, M., Teodorecu, R.I. (2022). Actual approaches for the craft beer fermentations. *Scientific Papers. Série B. Horticulture*, 66, 1.

Iturritxa, E., Hill, A.E., Torija, M.J. (2023). Profiling potential brewing yeast from forest and vineyard ecosystems. *International Journal of Food Microbiology*, 394, 110187.

Jackowski, M., Czepiela, W., Hampf, L., Zuczkowski, W., Dymkowski, T., Trusek, A. (2023). Comparison of two commercially available strains, *Saccharomycodes ludwigii* and *Torulaspora delbrueckii*, for the production of low-alcohol beer. *Beverages*, 9(3), 66.

Kato, T. and Takahashi, T. (2022). Studies on the genetic characteristics of the yeasts *Saccharomyces*: A review. *Journal of the American Society of Brewing Chemists*, 81(2), 1–12.

Kayadelen, F., Agirman, B., Jolly, N.P., Erten, H. (2023). The influence of *Torulaspora delbrueckii* on beer fermentation. *FEMS Yeast Research*, 23, 1–3.

Lin, C.L., Petersen, M.A., Mauch, A., Cottlieb, A. (2022). Towards lager beer aroma improvement via selective amino acid release by proteases during mashing. *Journal of the Institute of Brewing*, 128, 15–21.

Maiorella, B., Blanch, H.W., Wilke, C.R., (1983). By-product inhibition effects on ethanolic fermentation by *Saccharomyces cerevisiae*. *Biotechnology and Bioengineering*, 25(1), 103–121.

Manshin, D. (2022). Comparison of the yeast *Saccharomyces cerevisiae var. boulardii* and top-fermenting brewing yeast strains during the fermentation of model nutrient media and beer wort. *Agronomy Research*, 20(3), 625–636.

Millet, P. (2019). Bières de table techniques particulières de fabrication [Online]. Available at: https://www.btobeer.com/themes-conseils-techniques-bieres-brasseries/zythologie/bieres-de-table-techniques-particuliere-de-fabrication.

Nieto-Sarabia, V.L., Melgar-Lalanne, G., Ballinas-Cesatti, C.B., García-García, F.A., Jose-Salazar, J.A., Flores-Ortiz, C.M., Cristiani-Urbina, E., Morales-Barrera L. (2023). Brewing a craft Belgian-style pale ale using *Pichia kudriavzevii* 4A as a starter culture. *Microorganisms*, 11, 977.

Paszkot, J., Gasiński, A., Kawa-Rygielska, J. (2023). Evaluation of volatile compound profiles and sensory properties of dark and pale beers fermented by different strains of brewing yeast. *Scientific Reports*, 13, 6725. doi: 10.1038/s41598-023-33246-4.

Riess, J. (2012). Intensification de la brique “fermentation alcoolique” de substrats betteraviers pour la production d’éthanol. PhD Thesis, Institut National Polytechnique de Toulouse, Toulouse.

de Ruijter, J.C., Aisala, H., Jokinen, I., Krogerus, K., Rischer, H., Toivari, M. (2023). Production and sensory analysis of grape flavoured beer by co-fermentation of an industrial and a genetically modified laboratory yeast strain. *European Food Research and Technology*, 249, 1991–2000.

Thompson, A. and Dawson, P. (2023). Historical perspective and current challenges for microbreweries on bacterial spoilage of beer. *European Journal of Agriculture and Food Sciences*, 5(1), 80–84.

7

Production of Biomass and Bioactives by Microalgae

Hatem BEN OUADA and Jihène AMMAR

Laboratoire de Biotechnologie Bleu et Bioproduits Aquatiques (B3 Aqua), Institut National des Sciences et Technologies de la Mer (INSTM), Monastir Annex, Tunisia

7.1. Introduction

The autotrophic organisms known as microalgae derive their energy from oxygenic photosynthesis using water as an electron donor. However, under specific environmental conditions, such as high concentrations of dissolved organic carbon in the water, some microalgal species have the ability to become heterotrophic or photoheterotrophic (Perez-Garcia and Bashan 2015).

Microalgae proliferate in various environments, including aquatic, terrestrial, and even extreme conditions such as glaciers and arid deserts, as well as on various substrates such as soil, rocks, trees and even architectural buildings. Some species thrive in glacial meltwater or snow (Dittami et al. 2017). This adaptability is the result of morphological and genetic polymorphism, illustrated by a significant diversity comprising nearly 144,313 distinct species (Heimann and Huerlimann 2015), making their taxonomic classification complex. Advances in molecular phylogeny have contributed to a better understanding of these organisms and enabled constant updates in their classification.

In a global context aimed at reducing the use of chemical additives, microalgae production represents a high-potential alternative in various industries, including the food, cosmetics and health sectors. Their production offers a number of advantages

Bioreactor Implementation in the Agro-Food Industries,
coordinated by Mohamed GHOUL.

over terrestrial crops, including a reduced ecological footprint, more efficient use of resources and greater flexibility in the design of the production system. The use of microalgae biomass is particularly attractive because of its high productivity on non-arable and non-harvestable land. Microalgae can also adapt to a variety of environmental conditions thanks to the production of metabolites of interest, such as dietary supplements. Their benefits include improving the immune system, reducing viral infections, prebiotic and probiotic effects, increasing hemoglobin concentrations and reducing blood sugar levels (Caporgno and Mathys 2018; Kusmayadi et al. 2021).

Today, microalgae are exploited in the form of dry biomass (whole microalgae) or extracts, in several market segments, ranging from cosmetics to biofuels, fine chemicals, pharmaceuticals, foodstuffs, animal feeds, aquaculture, anaerobic digestion and pollution control.

Many compounds characterized by biological activities that are exploitable in the food and non-food sectors have already been used. However, microalgae remain one of the least studied groups of organisms in the world. Indeed, despite the large number of microalgae species recorded, only a few hundred have been studied in-depth, of which only a dozen are currently produced on a large scale. Guedes et al. (2011) estimated that around 97% of microalgae compounds have yet to be isolated and/or defined.

The aim of this summary is to describe the state of the art in the production of microalgae, as well as associated metabolites of interest, in the agri-food and biotechnology fields.

7.2. Biomass production from microalgae

In 2020, global microalgae production was estimated at around 10–15 kt of dry matter (Hachicha et al. 2022). This production is mainly attributed to three species, namely, *Arthrospira platensis* (5–7 kt/year), *Chlorella* sp. (3–4 kt/year) and *Dunaliella salina* (1–2 kt/year). Together, these three species account for 95% of total production (Show 2022). Other species, such as *Aphanizomenon flosaquae* (cyanobacteria) contribute 0.5 kt/year. Microalgae with the ability to produce specific metabolites, such as *Haematococcus pluvialis* (Chlorophyte), *Crypthecodinium cohnii* (Rhodophyte), *Shizochytrium* (diatom), *Nannochloropsis* (chlorophyte), *Odontella aurita*, *Porphyridium*, *Pheodactylum*, *Nitzschia*, and *Crypthecodinium*, contribute less than 1 kt/year (Mata et al. 2010; Show 2022).

Asia is the world's leading producer of microalgae, accounting for around 50% of total production. Other major producing countries include the United States, Chile, Argentina and Australia. In Europe, Germany and the Netherlands stand out as the leading producers, generating around 50 tons per year (Verdelho Viera 2015). In Mediterranean countries, information on the production, processing and marketing of microalgae and their derivatives remains scarce and inconsistent, particularly in southern nations. Rastoin (2017) nevertheless lists 72 companies that are active in the production of microalgae and their derivatives in nine Mediterranean countries: 50 within five European countries (Spain 15, France 21, Greece 1, Italy 6 and Portugal 7), as well as 22 companies in four southern countries (Isreal 10, Morocco 4, Tunisia 3 and Turkey 5). There is a proven presence in Egypt, and a probability that it exists in other countries such as Lebanon and Jordan.

In the industrial sector, microalgae are predominantly used in the form of dry biomass (whole microalgae). This biomass is mainly used as a food supplement in human nutrition (representing 74% by volume, 50% of which is for spirulina) and also in animal feed (25%, mainly for aquaculture). According to Market (2021a), the global microalgae market was valued at USD 977.3 million in 2020 and is expected to reach USD 1,485.1 million by 2028, registering a CAGR of 5.4%. According to another estimate (Polaris Market Research 2022), the global microalgae market was valued at USD 955.1 million in 2021 and is expected to reach USD 1,564.9 million by the end of 2030, at a CAGR of 5.7%. Overall, the microalgae market is expected to continue to grow in the years ahead, driven by increasing demand from a variety of industries and research and development efforts aimed at improving production profitability, as well as processing and creating high value-added products.

Indeed, microalgae contain a protein content in excess of 50% of their dry mass, with a high profile of amino acids that are essential for the body (Wang et al. 2021). In addition, various beneficial compounds such as antioxidants, vitamins and polyunsaturated fatty acids are present in high quantities. However, the use of microalgae as a dietary source of protein is still underdeveloped. Their high production costs compared to common plant protein sources (such as soy, lupin), as well as existing value chains and consumer acceptance in general, are the main constraints (Henchion et al. 2017).

Several types of functional products containing enriched microalgae have appeared in Europe since the 1990s. The most common are microalgae biomasses enriched with trace elements. The term "trace elements" or "micronutrients" refers to mineral salts that are present in the body in small quantities (less than 1 mg/kg body weight), but which are essential for all metabolic activities. Their absence or deficiency can lead to major disruptions in the metabolic processes that affect human health. The trace elements essential to life are divided into two categories according to their risk of deficiency: trace elements with a proven deficiency risk in humans, such as iodine, iron, copper, zinc, selenium, chromium, molybdenum and boron, and those with an unproven or low deficiency risk, such as manganese, silicon, vanadium, nickel, tin and arsenic.

Trace element enrichment is based on being soaked in a medium containing concentrations of trace element salts that are in excess of the algae's requirements, thus promoting its biosorption and metabolization capacity. Puyfoulhoux et al. (2001) describe a process for obtaining iron-enriched spirulina by adding $FeCl_3$ to the culture medium. Similarly, Cases et al. (1999, 2002) describe the production of selenium-enriched spirulina by the addition of sodium selenite. Ammar and Ben Ouada (2023) have described a rapid process based on the principle of osmoregulation, enabling enrichment rates in iron, zinc, chromium, iodine, manganese, copper, selenium and silicon to be achieved that exceed native biomass contents by 7, 1,155, 124, 214, 605, 290, 1,300 and over 15,000 times, respectively. Work by Nalimova et al. (2005) shows that iron metabolized by spirulina is much better assimilated than iron extracted from beef. Similarly, the bioavailability of metabolized zinc is better than that of zinc sulfate, and that of selenium is better than that of sodium selenite (Cases et al. 2002).

The market for microalgae enriched with trace elements is booming, especially with the current trend in the food industry, as well as in clinical practice (Schümann et al. 2007), using trace elements complexed with organic molecules prior to their incorporation into food and/or therapeutic products. Several types of complexing agents have appeared on the market in previous decades, including bovine hemoglobin and bovine lactoferrin for complexing iron, as well as horsetail powder (a wild plant) for transporting selenium in the body. However, this type of mechanical complexation is far from guaranteeing the performance of foods rich in (or enriched with) trace elements (Nalimova et al. 2005).

Enriched microalgae are mixed with biomass and sold on health food markets in compliance with recommended daily rations or used as pet food. Their high price limits their widespread use in livestock feed.

7.3. The main bioactives derived from microalgae

We are experiencing a resurging interest in bioactives derived from microalgae at the moment. There is currently a strong trend toward substituting synthetic components with natural products. This trend can be explained, on the one hand, by consumer demands in terms of health standards, ecology and respect for the environment and, on the other hand, by changes in legislation that favor natural products over synthetic ones (Pagliaro 2019).

The main bioactives derived from microalgae and available on the market are summarized in Table 7.1.

7.3.1. *Pigments*

Pigments are the most commercialized bioactives on the microalgal market. Microalgal pigments generally fall into three separate chemical categories:

– The tetraterpene or carotenoid group, containing substances with a yellow to red color, is made up of carotenes, which are hydrocarbons and xanthophylls, and which differ by the presence of oxygen in their molecules.

– The porphyrin group is made up of chlorophylls, which have a green color, and phycobiliproteins, which have a blue or red color. The latter are specific to blue-green algae (cyanobacteria), red-green algae (rhodophyceae) and cryptophyceae.

– Extracellular pigments are yellow to brown photoprotective pigments. They are represented by scytonemin and mycosporins.

7.3.1.1. *Carotenoids*

Microalgae are a particularly rich source of carotenoids. Some species contain up to 10 times more carotenoids than other plant sources, such as carrots, sweet potatoes and spinach (Zia-Ul-Haq 2021).

The market for algal carotenoids is dominated by beta-carotene, astaxanthin and lutein. Other carotenoids such as lycopene, zeaxanthin, fucoxanthin and canthaxanthin are much less widespread, but their economic potential remains promising (Novoveská et al. 2019).

7.3.1.1.1. β-Carotene

With a yellow-orange color, beta-carotene is the most widespread pigment on the microalgae market. It is a polyene composed of 40 carbon atoms, with the ends formed by two unsaturated rings.

The main source of β-carotene in microalgae is the chlorophyte *Dunaliella salina*, capable of accumulating β-carotene under conditions of salt, light, thermal stress and mineral nitrogen deficiency (Ben Amotz 1981). Growing green *D. salina* cells only contain 0.3% β-carotene, a level similar to that found in plants and other algae. Under stress conditions, β-carotene levels can reach up to 14% of dry weight (Borowitzka 1984).

β-Carotene is used as an ingredient in multivitamin supplements, as an additive in cosmetic preparations, as a colorant in animal feed and, above all, as an antioxidant in the human diet. It serves as a precursor to retinol (vitamin A), an extremely important molecule for vision, whose deficiency leads to growth problems and destruction of the intestinal epithelium (Krinsky 1989). This pigment also reduces cellular levels of oxidized DNA, which is implicated in cancer development (Mathews-Roth and Krinsky 1985).

The isomeric composition of *Dunaliella*'s natural β-carotene offers the benefits required for health products. It consists of two stereoisomers, all-trans and 9-cis, with more or less equal proportions, unlike chemically synthesized products that are solely composed of all-trans isomers (Harari et al. 2013). This composition results in superior bioavailability and health-promoting properties, justifying the commercial interest in *Dunaliella* carotenoids.

Dunaliella's natural β-carotene is now exploited in many markets through three different categories (Ben Amotz 2004): beta-carotene extracts, *Dunaliella* powder for human use and dried *Dunaliella* for animal feed.

– Purified β-carotene is mainly sold in vegetable oil at a concentration of 1–20%, for coloring various food products, as well as for personal use in soft gels, with a dose of around 5 mg of β-carotene per gel. Oils containing 75% β-carotene mixed with other carotenoids, mainly lutein, neoxanthin, zeaxanthin, violaxanthin, cryptoxanthin and alpha-carotene, are also available on the market under the name "carotenoid mixture". Currently, several formulations of natural β-carotene are available and distributed in health food stores, as well as in the vitamin, health food and dietary supplement market sectors.

– The second category comprises a range of low-salt dried *Dunaliella* powders. Harvested, processed, washed and dried algae are marketed as tablets or capsules containing between 3 and 20 mg of carotenoid mixture per unit. The tablets are coated with polysaccharides and packaged in capsules, then individually wrapped in aluminum/polyethylene blisters to prevent oxidation. Dried *Dunaliella* is widely marketed in the Middle East, where consumers are familiar with the use and consumption of algae.

– The third category comprises a range of *Dunaliella* that has been harvested and dried raw (with kosher salt). The algae powder contains around 2% β-carotene and a variable concentration of salt. This powder is vacuum-packed in aluminum/ polyethylene bags, intended for food coloring and provitamin A use in cattle, poultry, fish, shrimp, etc.

The market for carotenoids derived from microalgae is set to grow significantly over the coming years, driven by increasing demand for natural and sustainable sources in a variety of industries, including food, cosmetics and pharmaceuticals. It is estimated to reach USD 258 million by 2027, with a compound annual growth rate (CAGR) of 4.5% for the forecast period (Markets and Markets 2021). However, one obstacle remains for the marketing of algal β-carotene in general. Its price is significantly higher than that of its synthetic equivalent. It can reach up to EUR 700 per kilogram, compared with a market value for the synthetic product estimated at half that amount (Guedes et al. 2011). At present, the synthetic form covers 97–98% of the β-carotene market, while the remaining 2–3% comes from natural bioresources such as fruits, vegetables, algae, fungi, bacteria and microalgae (Marino et al. 2020).

7.3.1.1.2. Astaxanthin

Astaxanthin is a xanthophyll carotenoid pigment used as a dietary supplement and is responsible for the reddish-pink color of farmed salmon flesh. Its chemical structure is characterized by a central carbon–carbon double bond and two oxygen-containing functional groups at either end. Its molecular formula is C40H52O4 and its molecular weight is 596.85 g/mol (Meléndez-Martínez et al. 2007).

Microalgae are considered one of the most promising sources of astaxanthin, due to their high productivity, scalability and relatively low production costs. Although astaxanthin can also be synthesized by other organisms such as bacteria, certain fungi and shrimp, the quantities accumulated by the microalga *Haematococcus pluvialis* can reach 40 g/kg of dry biomass, exceeding those obtained from the sources mentioned above by a factor of 5 to more than 20 times (Boussiba and Vonshak 1991).

Currently, *Haematococcus pluvialis* is the most widely used microalgae for commercial astaxanthin production, accounting for over 90% of the global market share. Other potential sources are currently being explored, including the microalgae

Chlorella zofingiensis and *Chlorococcum* spp. (Jannel et al. 2020). Astaxanthin accumulation in *Haematococcus* occurs during the transition from the vegetative state (green phase) to the steady phase (red phase). This accumulation is favored by stress factors, whether they be physical (temperature, light intensity) or chemical (pH, salinity, free radicals, etc.) (Camacho Kurmen et al. 2013).

Astaxanthin's benefits are linked to antibody production, anti-tumor and anti-inflammatory activities, inhibition of colon, bladder, liver, breast and oral cancers, reduced risk of diseases such as Parkinson's and Alzheimer's, and improved cardiovascular health (Mata et al. 2010).

The astaxanthin market has been developing rapidly, driven by growing demand for natural, sustainable sources of pigments. It is expected to reach USD 1.1 billion by 2027, with a CAGR of 8.4% (Market and Market 2021). Overall, astaxanthin production from microalgae represents a promising and sustainable approach to meeting the growing demand for this high-potential pigment.

7.3.1.1.3. Lutein

Lutein is a xanthophyll carotenoid pigment with a linear structure of 40 carbon atoms and 56 hydrogen atoms. It has two main functional groups: a long hydrocarbon chain and a polyene chain of conjugated double bonds, giving it its yellow color and antioxidant properties.

Several microalgae (*Muriellopsis*, *Desmodesmus* sp. and *Scenedesmus*) are considered to be potential sources of lutein (Zheng et al. 2022). Their content of this pigment, estimated at 0.5–1.2% of dry weight, is much higher than that of the usual sources (marigold petals, green leafy vegetables and certain fruits), whose content does not exceed 0.05% of dry weight (Fernandez-Sevilla et al. 2010). Lutein is mainly accumulated autotrophically, but also mixotrophically (Chen et al. 2019) and heterotrophically (*Chlorella* and *Dunaliella*) (Nagarajan et al. 2018).

Lutein's beneficial effects on human health are attributed to its strong antioxidant and anti-inflammatory activity. Lutein has been widely used in eye health to palliate age-related macular degeneration, cataracts, retinal nerve disease, Alzheimer's disease and diabetic retinopathy (Steiner et al. 2018).

The lutein market is growing rapidly. It was valued at around USD 250 million in 2020, and it is expected to reach USD 400 million by 2025. The share of microalgae-based production is making a significant contribution to this growth.

7.3.1.2. *Phycobiliproteins*

Blue-green algae, like red algae and cryptophyceae (flagellate algae), are distinguished from the plant kingdom by their synthesis of phycobilins. Phycobilins differ from animal bilins in that they are bound to proteins, hence the name phycobiliproteins. Phycobilins belong to the family of tetrapyrrole pigments, which also includes bilins, porphyrins such as chlorophyll and hemoglobin. They are distinguished by the arrangement of their four pyrrolic rings in an open linear chain and by their union by three carbon bridges instead of four (Marizcurrena et al. 2019).

The main phycobiliprotein produced on a large scale is phycocyanin (Spolaore et al. 2006), which holds this monopoly because of its unique blue color.

The list of natural colorants available on the market suffers from a considerable lack of blue colorants. The only examples are chamazulene (extracted from chamomile flowers) and indigo (derived from the fermentation of indigo leaves). However, the availability of these two dyes does not meet market needs, due to low extraction yields of no more than 0.1% of dry weight (Buchweitz 2016). Phycocyanin sold on the world market is exclusively extracted from the cyanobacterium *Arthrospira platensis* (spirulina). Spirulina is microalgae that is produced on a large scale and has the advantage of an exceptional phycocyanin content of up to 20% of its dry weight (Fabre et al. 2022). The highest rates of synthesis occur with nitrogen levels exceeding those administered in industrial spirulina cultures, and at lower light intensities than those optimal for growth (Wicaksono et al. 2019).

Phycocyanin's antioxidant and fluorescent properties give it a variety of pharmaceutical and medical applications (Erikson 2008). Its role as an immune system stimulant and its anti-cancer properties, which has been known for decades (Jiang et al. 2017), make it a compound of great interest to the parapharmaceutical industries. This pigment also serves as a biochemical tracer in microscopy and flow cytometry because of its fluorescence (Herrera et al. 1989). Its naturalness, aqueous extraction without organic solvents, therapeutic properties and unique blue color makes it an ideal candidate for the food market, in line with European and American standards.

The industrial use of phycocyanin has long been inconvenienced by problems of preservation and stability. Extraction in aqueous solvents encourages the growth of bacteria and fungi, limiting shelf life. Phycocyanin is also highly photosensitive, leading to rapid degradation of the molecule, even under low light conditions

(Fitzsimons 2010). Currently, industrial applications for phycocyanin are limited to a few food products (marzipans, sweet juices, ice creams, sorbets and occasionally chocolates) and a few cosmetic formulations (creams and gels). Market sales of phycocyanin are monopolized by a few (mainly Japanese) companies and, at most, cover only 2% of the US market for natural food colorants (Griffiths et al. 2016).

Various phycocyanin stabilization processes have been developed in recent years. The most suitable are those using food acids such as ascorbic acid (Minard 2005; Kannaujiya and Sinha 2016) or salicylic acid, or the incorporation of the molecule into a polyol such as glycerol (Ben Ouada and Ammar 2016).

Resolving stability issues related to phycocyanin has enabled the global market to flourish in recent years. Research and Markets (2021b) forecasted that the global phycocyanin market will reach a value of USD 409.8 million by 2030.

7.3.1.3. *Extracellular pigments*

Like bacteria and fungi, microalgae secrete various low molecular weight substances into their environment, such as toxins (Aráoz et al. 2005), antimicrobial, antifungal and anti-algal substances (Volk and Furkert 2006), as well as phytohormones (Zaccaro et al. 2006). However, microalgae are distinguished by their ability to synthesize photoprotective substances. The most widely available are known as scytonemins and micosporin-like amino acids (MAAs).

7.3.1.3.1. Scytonemin

Scytonemin is a fat-soluble, yellow-brown pigment found in the extracellular polysaccharides of over 300 species of cyanobacteria (Sinha and Häder 2008). It is a 544 Da dimer made up of two sub-units: one being indolic and one being phenolic. Scytonemin can exist in a yellow oxidized form, fuscochlorin, and in a red reduced form, fuscorhodin. The existence of one form or the other mostly depends on redox and acid–base conditions during the extraction process (Rastogi et al. 2015).

It is an absorbing pigment at 370 nm, known for its ability to protect cells from harmful UV rays. The UV-protective effect of scytonemin was first observed in land-based cyanobacteria by Nägeli and Gattungen in 1849 (Sinha and Häder 2008). Scytonemin can prevent up to 90% of solar UV radiation from entering the cell (Garcia-Pichel and Castenholz 1991). In addition, this extracellular pigment has a high capacity to scavenge free radicals and has great stability against various abiotic factors such as UV-B (0.78 W m^{-2}), heat (60°C) and oxidation (Rastogi et al. 2015). Scytonemin has great potential for application in several fields, with interesting anti-inflammatory and anti-proliferative activities (Stevenson et al. 2002).

Scytonemin production is currently limited to extraction from natural sources, making it difficult and costly to exploit. However, there is growing interest in producing scytonemin using recombinant DNA technology, which involves inserting the gene responsible for scytonemin production into a host organism (Xiang et al. 2021).

Although there is currently no significant market for scytonemin, there is potential for growth in the future as demand for natural agents increases (Derikvand et al. 2017).

7.3.1.3.2. Mycosporins

Mycosporins generally consist of a cyclic cyclohexenone or cyclohexenimine core, with one or more side chains of amino acids such as glycine, serine, valine and threonine. This is why they are commonly referred to as MAAs. They are water-soluble molecules with a slightly yellowish color and a molecular weight ranging from 188 to 1,050 Da (Wada et al. 2015). MAAs were initially observed in terrestrial fungi, but were first described in the cyanobacterium *Gloeocapsa* sp. (Garcia-Pichel et al. 1993).

Microalgae are proving to be a particularly promising source of mycosporins, being able to produce large quantities of these compounds in response to UV exposure (Rastogi and Incharoensakdi 2014).

The production of mycosporins by microalgae is still a relatively recent field of research, and the market for these compounds is still emerging. However, there is growing interest in the use of mycosporins in the food and cosmetics industries, which exploit their antioxidant activity. Overall, the market for mycosporins produced by microalgae is set to grow in the coming years as more targeted research is carried out into the production, structure and potential applications of these compounds.

7.3.2. *Extracellular polymeric substances*

Extracellular polymeric substances (EPS) are polymeric complexes composed of several high molecular weight fractions, ranging from 10 to 2,000 kg/mol (Babiak and Krzeminska 2021; Gongi et al. 2022). For the most part, they are heteropolysaccharides that integrate various monosaccharides, generally between 9 and 12, with the presence of several non-sugar substituents such as acyl groups, proteins, sulfate groups or uronic acids (Delattre et al. 2016; Gongi et al. 2022). Carbohydrates generally represent between 40% and 95% of the mass of EPS.

The most commonly found oses are D-arabinose, D-ribose, D-xylose for pentoses, and D-glucose, D-galactose, D-mannose, D-allose, L-rhamnose and L-fucose for hexoses (Haras 2005). Several enzymes may also be present in EPS, such as superoxide dismutase (SOD), catalase and ascorbate peroxidase (APX) (Gongi et al. 2022).

EPS are produced by several species of microalgae in response to various stress conditions. In common chlorophytic microalgae, EPS productivity reaches 0.38 g/L of culture (Halaj et al. 2018). Higher EPS contents have been reported in cyanobacteria (1.35–2.50 g/L; Soanen et al. 2016). Thermophilic species are known to produce large amounts of EPS as a means of adapting to extreme temperatures (Vu et al. 2009) with contents that can exceed 3 g/L of culture (Gongi et al. 2022).

Microalgal EPS belong to the hydrocolloid family, recognized for their thickening, gelling, stabilizing, emulsifying, binding and flocculating properties, as well as for their thermostability (Delattre et al. 2016; Xiao and Zheng 2016). Some microalgal EPS possess various biological activities, including being anti-inflammatory (Bae et al. 2006), immunomodulatory (Guzmán et al. 2003), antitumor (Raposo et al. 2014), antiviral (Huheihel et al. 2002), and antifungal and antibacterial (Gongi et al. 2022). Microalgal EPS have also been shown to possess antioxidant activity (Xiao and Zheng 2016) and metal-chelating properties (Gongi et al. 2022).

Microalgal EPS have been used as emulsifiers, thickeners, flocculants and gelling agents (Mishra et al. 2011; Han et al. 2014) for applications in food, pharmaceuticals, decantation, sludge dewatering, as well as metal removal during mining and industrial waste treatment (Abed et al. 2009). They have been recognized as a potential alternative to conventional chemical polymers, due to their biodegradability, biocompatibility and non-toxicity (Pereira et al. 2019).

Typically, over 8% of polymers on the market are manufactured from fossil resources. This proportion could rise to 20% by 2050 (Schneiderman and Hillmyer 2017). Due to growing concern about environmental damage and the exhaustibility of these resources, there has been a particular focus on the use of natural resources (Hong and Chen 2017). Common natural polymers are extracted from macroalgae (agar, carrageenan and alginate) and bacteria (dextran and xanthan gum) (Brunchi et al. 2016; Martău et al. 2019). The benefit of microalgal EPS is that they are recovered directly from the culture medium through simple centrifugation and/or ultrafiltration (De Philippis and Vincenzini 1998; Gongi et al. 2022).

Currently, the market for EPS produced by microalgae remains limited compared with other sources, mainly due to the structural complexity of EPS that present various functions, which can vary according to cultivation conditions and species. However, as a result of the growing demand for natural and sustainable ingredients, it is developing rapidly.

7.3.3. *Lipids*

Lipids produced by microalgae represent a promising source of sustainable and renewable energy, containing a wide range of bioactives. The market for microalgae-based lipids is developing rapidly, with applications in various industries such as:

– *Biofuels*: microalgae lipids can be converted into biodiesel, a renewable and environmentally friendly alternative to fossil fuels. Biodiesel produced from microalgae has a higher energy density and is more efficient than traditional sources of biodiesel (Culaba et al. 2020).

– *Food and beverages*: microalgae-based lipids are rich in omega-3 fatty acids, which offer numerous health benefits. They can be used as dietary supplements or incorporated into a variety of foods and beverages (Remize et al. 2021).

– *Cosmetics*: microalgae lipids possess antioxidant and anti-inflammatory properties, making them ideal ingredients for skin care and personal care products (De Luca et al. 2021).

– *Animal feed*: microalgae-based lipids can be used as feed supplements for livestock and aquaculture. They provide a source of essential fatty acids and other nutrients that are beneficial to animal health and growth (Soto-Sánchez et al. 2023).

7.3.4. *Polyunsaturated fatty acids*

Polyunsaturated fatty acids (PUFAs) are long-chain essential fatty acids containing more than one double bond. Microalgae are considered a promising source for the production of these fatty acids, particularly polyunsaturated fatty acids (Ohlrogge and Browse 1995). PUFAs account for 25%–60% of total fatty acids in microalgae (Caudales et al. 2000), but these proportions vary according to species and culture conditions (Leonardi et al. 2011).

The PUFAs of interest in chlorophyceae are mainly arachidonic acid (ARA; 20:4n-6), docosahexaenoic acid (DHA; 22:6n-3) and eicosapentaenoic acid (EPA, 20:5n-3) (Li et al. 2015; Bouzidi et al. 2020). In cyanobacteria, linoleic acid

(LA; C18:2n-6) and gamma-linolenic acid (GLA; C18:3n-6) are found as the major fatty acids (Hu et al. 2008).

PUFAs are involved in the modulation of numerous physiological processes such as cardiovascular, neurological and immune functions (Yashodhara et al. 2009). Long-chain omega-3 and omega-6 PUFAs are used to stimulate neurological and cardiac development, and contribute to the reduction of skin disease, rheumatism, coronary heart disease, hypertension, cancer and cholesterol (Mata et al. 2010). It is now known that omega-6 PUFAs, the main ones being LA and ARA, and their derivatives, have pro-inflammatory, pro-oxidant and pro-platelet-aggregating potential. Omega-3 PUFAs, such as EPA and DHA, have anti-arrhythmic, anti-inflammatory and antithrombotic effects (Schoenfeld 2005).

The current commercial source of long-chain PUFAs is fish oil, which is more competitive on the market than that derived from microalgae. However, problems of possible toxin accumulation in fish have been raised (Apt and Behrens 1999). The application of fish oil in food additives is also constrained by its typical fishy odor, unpleasant taste, low oxidation stability and the presence of mixed fatty acids. These disadvantages argue in favor of the use of algal PUFAs in food applications (Abril et al. 2003).

The market value of PUFAs from microalgae varies according to a number of factors, including the type of PUFA, production methods and application. PUFAs commonly produced from microalgae include EPA, DHA and ARA. Although some species have demonstrated the potential for industrial production of EPA, GLA and ARA, only DHA is currently economically competitive on the market (Sharma et al. 2020). Several species of microalgae are known to produce DHA. The most commonly used on a commercial scale are *Crypthecodinium cohnii*, *Schizochytrium* sp., *Nannochloropsis* sp. and *Isochrysis* sp.

DHA is currently on the market in the form of oils for infant formula, with DHA levels in the range of 40%–45% w/w, and virtually devoid of EPA. In addition, oils rich in DHA (35%–40%) and low levels of EPA are marketed for the food, beverage and supplement industries (Winwood 2013).

According to a report by Transparency Market Research (2019), the global market for EPA/DHA omega-3 ingredients was valued at USD 4.1 billion in 2019, and it is expected to grow at a CAGR of 7.5% from 2020 to 2027. Overall, the market for PUFAs from microalgae is expected to grow in the coming years, due to increasing demand for sustainable, plant-based sources of PUFAs in various industries.

7.4. Other microalgae potential

Research and screening of microalgae with industrial potential is still ongoing, targeting strains characterized by high productivity in terms of biomass or specific metabolites. In particular, thermophilic and thermotolerant microalgae are described as sources with high future potential. The species mentioned (*Synechococcus*, *Graesiella*, *Leptolyngbya*, *Halamphora*, *Gloecapsa*, etc.) offer advantages for industrial production, particularly in summer outdoor climatic conditions, where high temperatures inhibit the growth of common microalgae species. In addition, they are able to synthesize various high-value compounds such as HSPs (heat shock proteins) (Skjånes et al. 2021), extracellular polysaccharides (Gongi et al. 2021, 2022), various thermostable enzymes stabilizing protein and nucleic acid conformations (Rothschild and Mancinelli 2001), and unusual lipids forming highly impermeable membranes (Zili et al. 2015; Bouzidi et al. 2020).

Systematic research carried out since 1996 (Namikoshi and Rinehart 1996) has led to the isolation and identification of several bioactive compounds with potential applications in various fields. To date, over 600 natural products have been described from microalgae strains. Cyanobacteria such as *Anabaena*, *Nostoc*, *Microcystis*, *Lyngbya*, *Oscillatoria*, *Phormidium* and *Spirulina* produce a variety of high-value compounds, including cytotoxic (41%), antitumoral (13%), antiviral (4%), antimicrobial compounds (12%) and others (18%), including antimalarials, antimycotics, multidrug resistance reversers, herbicides and immunosuppressants (Burja et al. 2001). A new class of HIV inhibitors, called sulfonic acid, containing a glycolipid, has been isolated from cyanobacterial extracts. Mansour et al. (2011) reported that polysaccharides isolated from *Gloeocapsa turgidus* and *Synechococcus cedrorum* show strong antiviral activity against rabies and herpes 1 viruses. Extracellular polymers from *Aphanothece halophytica* show antiviral activity against influenza A (H1N1) virus, with 30% inhibition of pneumonia in infected mice (Zheng et al. 2006).

Microalgae are also a source of peptides with antimicrobial, antiviral and anticancer properties. Some of these peptides have been identified and characterized, generating growing interest in their potential therapeutic applications. Cyanovirin-N (CVN), a peptide isolated from cyanobacteria, inactivates HIV virus strains and inhibits cell-to-cell and virus-to-cell fusion (Patterson 1996). Several chlorophytes (*Chlorella*, *Dunaliella*, *Pavlova*) produce peptides with antihypertensive, antidiabetic, anti-UV, anti-osteoperotic, antibiotic and antiviral effects (Skjånes et al. 2021).

Bioactives	Type	Main species	Use	Sector
β-Carotene	Orange pigment, carotene, fat-soluble	*Dunaliella salina*	Colorant, food supplement, active ingredient	Cosmetics, food, parapharmacy
Astaxanthin	Orange-red pigment, Xanthophyll, fat-soluble	*Haematococcus pluvialis*, *Chlorella zofingiensis*, *Chlorococcum*	Colorant, food supplement, active ingredient	Medicine, food
Lutein	Yellow pigment, Xanthophyll, fat-soluble	*Muriellopsis*, *Desmodesmus sp*, *Scenedesmus*, *Chlorella*, *Dunaliella*	Colorant, food supplement, active ingredient	Medicine, eye health
Phycocyanin	Blue pigment, biliprotein, water-soluble	*Arthrospira platensis*	Colorant, food supplement, fluorescent tracer	Pharmacy, medicine, cosmetics, dietetics
Phycoerythrin	Red pigment, biliprotein, water-soluble	*Porphyridium* sp.	Colorant, fluorescent tracer	Pharmacy, medicine, cosmetics
Scytonemin	Extracellular pigment, yellow-brown, fat-soluble	Cyanobactéries	Sunscreen	Cosmetics and parapharmacy
Mycosporins	Extracellular pigment, yellowish, water-soluble	Cyanobactéries	Sunscreen	Cosmetics and parapharmacy
Extracellular polymeric substances (EPS)	Hetero-polysaccharides	*Gyrodinium impudicum*, *Porphyridium marinum*, *Grasiella*, *Arthrospira*, *Gloeocapsa*.	Emulsifiers, thickeners, flocculants, gelling agents, active ingredient	Medicine, pharmacy
Eicosa-pentaenoic acid (EPA)	Polyunsaturated fatty acids (omega 3)	*Monodus subterraneus*	Food supplements, active ingredients	Dietetics and food supplements
Docosahexaenoic acid (DHA)	Polyunsaturated fatty acids (omega 3)	*Crypthecodinium cohnii*, *Schizochytrium* sp., *Nannochloropsis* sp., *Isochrysis* sp.	Food supplements, active ingredients	Food, cosmetics, animal feed
Arachidonic acid (ARA)	Polyunsaturated fatty acids (omega 6)	*Euglena*, *Porphyridium*, *Parietochloris*	Food supplements, active ingredients	Food, cosmetics, animal feed

Table 7.1. *Main bioactives marketed from microalgae*

7.5. Conclusion

The aim of this summary is to provide a relevant and concise update on the current state of global production of microalgae and bioactives for food and biotechnology applications. Microalgae are promising sources of proteins, polyunsaturated fatty acids, minerals, carotenoids, colloidal substances, pigments, etc. They have the ability to generate a variety of bioactive compounds with beneficial health effects, such as antioxidant, anti-inflammatory, antimicrobial and antiviral properties.

All market studies tend to predict a substantial and sustained development of microalgae production in the short to medium term. Indeed, the growing volume of scientific and technological information points to great potential for large-scale production in the near future.

However, scaling up production remains a major challenge if commercial competitiveness is to be achieved. The microalgae sector is still subject, on the one hand, to uncompetitive business models that are based on fossil carbon and chemical synthesis; and, on the other hand, to the multiple traditional sources of recoverable biomass, with technological models and value chains that have stayed the same for several generations.

This being said, microalgae production is in line with the global objective of establishing sustainable consumption and production patterns, using natural resources, in a context of combating climate change, water scarcity, the fragility of rural areas and food insecurity. Using microalgae biomass within a localized and participatory bioeconomy should ensure the sustainability of the sector and its future development.

7.6. References

Abed, R., Dobretsov, S., Sudesh, K. (2009). Applications of cyanobacteria in biotechnology. *J. Appl. Microbiol.*, 106, 1–12.

Abril, R., Garrett, J., Zeller, S.G., Sander, W.J., Mast, R.W. (2003). Safety assessment of DHA-rich microalgae from Schizochytrium sp. Part V: Target animal safety/toxicity study in growing swine. *Regul. Toxicol. Pharm.*, 37, 73–82.

Ammar, J. and Ben Ouada, H. (2023). Traitement post culture de courte durée permettant un fort taux d'enrichissement de microalgues en oligoéléments. Patent, TN 2022/0205.

Apt, K.E. and Behrens, P.W. (1999). Commercial developments in microalgal biotechnology. *J. Phycol.*, 35, 215–226.

Aráoz, R., Nghiêm, H.O., Rippka, R., Palibroda, N., De Marsac, N.T., Herdman, M. (2005). Neurotoxins in axenic oscillatorian cyanobacteria: Coexistence of anatoxin-a and homoanatoxin-a determined by ligand-binding assay and GC/MS. *Microbiology*, 151(4), 1263–1273.

Babiak, W. and Krzemińska, I. (2021). Extracellular polymeric substances (EPS) as microalgal bioproducts: A review of factors affecting EPS synthesis and application in flocculation processes. *Energies*, 14(13), 4007.

Bae, S.Y., Yim, J.H., Lee, H.K., Pyo, S. (2006). Activation of murine peritoneal macrophages by sulfated exopolysaccharide from marine microalga *Gyrodinium impudicum* (strain KG03): Involvement of the NF-κB and JNK pathway. *Int. Immunopharmacol*, 6, 473–484.

Ben Amotz, A. (1981). Glycerol and ß-carotene metabolism in the halotolerant alga *Dunaliella*: A model system for biosolar energy conversion. *Trends in Biochemical Sciences*, 6, 297–299.

Ben Amotz, A. (2004). Industrial production of microalgal cell-mass and secondary products – Major industrial species: Dunliella. In *Handbook of Microalgal Culture: Biotechnology and Applied Phycology*, Richmond, A. (ed.). Blackwell Science, Oxford, 273–280.

Ben Ouada, H. and Ammar, J. (2016). Method for producing a stable precipitate enriched in phycobiliproteins. Patent, WO2016030643A1.

Borowitzka, M.A. (1984). Calcification in aquatic plants. *Plant, Cell and Environment*, 7(6), 457–466.

Boussiba, S. and Vonshak, A. (1991). Astaxanthin accumulation in the green alga *Haematococcus pluvialis*. *Plant and Cell Physiology*, 32(7), 1077–1082.

Bouzidi, N., Zili, F., García-Maroto, F., Alonso, D.L., Ben Ouada, H. (2020). Impact of temperature and growth phases on lipid composition and fatty acid profile of a thermophilic Bacillariophyta strain related to the genus *Halamphora* from north-eastern Tunisia. *Journal of the Marine Biological*, 100(4), 529–536.

Brunchi, C.E., Bercea, M., Morariu, S., Dascalu, M. (2016). Some properties of xanthan gum in aqueous solutions: Effect of temperature and pH. *J. Polym. Res.*, 23, 126.

Buchweitz, M. (2016). Natural solutions for blue colors in food. In *Handbook on Natural Pigments in Food and Beverages: Industrial Applications for Improving Food Color*, Carle, R. and Schweiggert, R.M. (eds). Woodhead Publishing, Cambridge, 355–384.

Burja, A.M., Banaigs, B., Abou-Mansour, E., Burgess, J.G., Wright, P.C. (2001). Marine cyanobacteria – A prolific source of natural products. *Tetrahedron*, 57(46), 9347–9377.

Camacho Kurmen, J.E., Gonzalez, G., Klotz, B. (2013). Astaxanthin production in *Haematococcus pluvialis* under different stress conditions. *Nova*, 11(19), 94–104.

Caporgno, M.P. and Mathys, A. (2018). Trends in microalgae incorporation into innovative food products with potential health benefits. *Frontiers in Nutrition*, 5, 58.

Cases, J., Puiga, M., Caporiccioa, B., Baroux, B., Baccouc, J.C., Besancona, P., Rouanet, J.-M. (1999). Glutathione-related enzymic activities in rats receiving high cholesterol or standard diets supplements with two forms of selenium. *Food Chem.*, 65, 207–211.

Cases, J., Agnieszka Wysocka, I., Caporiccio, B., Jouy, N., Besançon, P., Szpunar, J., Rouanet, J.-M. (2002). Assessment of selenium bioavailability from high-selenium spirulina subfractions in selenium-deficients rats. *J. Agric. Food Chem.*, 50, 3867–3873.

Caudales, R., Wells, J.M., Butterfield, J.E. (2000). Cellular fatty acid composition of cyanobacteria assigned to subsection II, order Pleurocapsales. *International Journal of Systematic and Evolutionary Microbiology*, 50(3), 1029–1034.

Chen, J.-H., Chen, C.-Y., Hasunuma, T., Kondo, A., Chang, C.-H., Ng, I.-S. (2019). Enhancing lutein production with mixotrophic cultivation of *Chlorella sorokiniana* MB-1-M12 using different bioprocess operation strategies. *Bioresource Technology*, 27, 17–25.

Culaba, A.B., Ubando, A.T., Phoebe, M.L.C., Chen, W.-H., Chang, J.-S. (2020). Biofuel from microalgae: Sustainable pathways. *Sustainability*, 12(19), 8009.

De Luca, M., Pappalardo, I., Limongi, A.R., Viviano, E., Radice, R.P., Todisco, S., Martelli, G., Infantino, V., Vassallo, A. (2021). Lipids from microalgae for cosmetic applications. *Cosmetics*, 8, 52.

De Philippis, R. and Vincenzini, M. (1998). Exocellular polysaccharides from cyanobacteria and their possible applications. *FEMS Microbiology*, 22, 151–175.

Delattre, C., Pierre, G., Laroche, C., Michaud, P. (2016). Production, extraction and characterization of microalgal and cyanobacterial exopolysaccharides. *Biotechnology Advances*, 34(7), 1159–1179.

Derikvand, P., Llewellyn, C.A., Purton, S. (2017). Cyanobacterial metabolites as a source of sunscreens and moisturizers: A comparison with current synthetic compounds. *Eur. J. Phycol.*, 52, 43–56.

Dittami, S.M., Heesch, S., Olsen, J.L., Collén, J. (2017). Transitions between marine and freshwater environments provide new clues about the origins of multicellular plants and algae. *J. Phycol.*, 53, 731–745.

Erikson, N.T. (2008). Production of phycocyanin – A pigment with applications in biology, biotechnology, foods and medicine. *Applied Microbiology and Biotechnology*, 80, 1–14.

Fabre, J.F., Niangoran, N.U.F., Gaignard, C., Buso, D., Mouloungui, Z., Valentin, R. (2022). Extraction, purification and stability of C-phycocyanin from *Arthrospira platensis*. *European Food Research and Technology*, 248, 1583–1599.

Fernández-Sevilla, J.M., Acién Fernández, F.G., Molina Grima, E. (2010). Biotechnological production of lutein and its applications. *Appl. Microbiol. Biotechnol.*, 86(1), 27–40.

Fitzsimons, T.R. (2010). Harnessing solar energy using photosynthetic and organic pigments. PhD Thesis, University of Central Oklahoma, Oklahoma.

Garcia-Pichel, F. and Castenholz, R.W. (1991). Characterization and biological implications of scytonemin, a cyanobacterial sheath pigment. *Journal of Phycology*, 27, 395–409.

Garcia-Pichel, F., Wingard, C., Castenholz, R.W. (1993). Evidence regarding the UV sunscreen role of a mycosporine-like compound in the cyanobacterium *Gloeocapsa* sp. *Appl. Environ. Microbiol.*, 59, 170–176.

Gongi, W., Cordeiro, N., Gomez Pinchetti, J.L., Ben Ouada, H. (2021). Production of exopolymer substances from the thermophilic chlorophyte *Graesiella*: Industrial and ecological applications. *Journal of Applied Phycology*, 33, 343–356.

Gongi, W., Gomez Pinchetti, J.L., Cordeiro, N., Ben Ouada, H. (2022). Extracellular polymeric substances produced by the thermophilic cyanobacterium *Gloeocapsa gelatinosa*: Characterization and assessment of their antioxidant and metal-chelating activities. *Marine Drugs*, 20(4), 227.

Griffiths, M., Harrison, S.T.L., Smit, M., Maharajh, D. (2016). Major commercial products from micro and macroalgae. In *Algae Biotechnology: Green Energy and Technology*, Bux, F. and Chisti, Y. (eds). Springer International Publishing, Cham, 269–300.

Guedes, A.C., Amaro, H.M., Malcata, F.X. (2011). Microalgae as sources of carotenoids. *Marine Drugs*, 9, 625–644.

Guzmán, S., Gato, A., Lamela, M., Freire-Garabal, M., Calleja, J.M. (2003). Anti-inflammatory and immunomodulatory activities of polysaccharide from *Chlorella stigmatophora* and *Phaeodactylum tricornutum*: Anti-inflammatory and immunodulatory activities. *Phytother. Res.*, 17, 665–670.

Hachicha, R., Elleuch, F., Ben Hlima, H., Dubessay, P., de Baynast, H., Delattre, C., Fendri, I. (2022). Biomolecules from microalgae and cyanobacteria: Applications and market survey. *Applied Science*, 12(4), 1924.

Halaj, M., Chvalova, B., Cepak, V., Lukavsk, Y.J., Capek, P. (2018). Searching for microalgal species producing extracellular biopolymers. *Chem. Pap.*, 72, 2673–2678.

Han, P., Sun, Y., Wu, X., Yuan, Y., Dai, Y., Jia, S. (2014). Emulsifying, flocculating, and physicochemical properties of exopolysaccharide produced by cyanobacterium *Nostoc flagelliforme*. *Appl. Biochem. Biotechnol.*, 172, 36–49.

Harari, A., Abecassis, R., Relevi, N., Levi, Z., Ben-Amotz, A., Kamari, Y., Harats, D., Shaish, A. (2013). Prevention of atherosclerosis progression by 9-cis-β-carotene rich alga Dunaliella in apoE-deficient mice. *BioMed Research International*, 169517.

Haras, D. (2005). Biofilms et altérations des matériaux : de l'analyse du phénomène aux stratégies de prévention. *Matériaux et Techniques*, 93, s-27.

Heimann, K. and Huerlimann, R. (2015). Microalgal classification: Major classes and genera of commercial microalgal species. In *Handbook of Marine Microalgae: Biotechnology Advances*, Kim, S.-K. (ed.). Academic Press, London, 25–41.

Henchion, M., Hayes, M., Mullen, A., Fenelon, M., Tiwari, B. (2017). Future protein supply and demand: Strategies and factors influencing a sustainable equilibrium. *Food*, 6, 53.

Herrera, A., Boussiba, S., Napoleone, V., Hohlberg, A. (1989). Recovery of c-phycocyanin from the cyanobacterium *Spirulina maxima. Journal of Applied Phycology*, 1, 325–331.

Hong, M. and Chen, E.Y.X. (2017). Chemically recyclable polymers: A circular economy approach to sustainability. *Green Chem.*, 19, 3692–3706.

Hu, Q., Sommerfeld, M., Jarvis, E., Ghirardi, M., Posewitz, M., Seibert, M., Darzins, A. (2008). Microalgal triacylglycerols as feedstocks for biofuel production: Perspectives and advances. *The Plant Journal*, 54(4), 621–639.

Huheihel, M., Ishanu, V., Jacov, T., Arad, S. (2002). Activity of *Porphyridium* sp. polysaccharide against herpes simplex viruses in vitro and in vivo. *J. Biochem. Biophys. Meth.*, 50, 189–200.

Jannel, S., Caro, Y., Bermudes, M., Petit, T. (2020). Novel insights into the biotechnological production of *Haematococcus pluvialis* derived astaxanthin: Advances and key challenges to allow its industrial use as novel food ingredient. *Journal of Marine Science and Engineering*, 8(10), 789.

Jiang, L., Wang, Y., Yin, Q., Liu, G., Liu, H., Huang, Y., Li, B. (2017). Phycocyanin: A potential drug for cancer treatment. *J. Cancer*, 8(17), 3416–3429.

Kannaujiya, V.K. and Sinha, R.P. (2016). Thermokinetic stability of phycocyanin and phycoerythrin in food-grade preservatives. *Journal of Applied Phycology*, 28, 1063–1070.

Krinsky, N.I. (1989). Antioxidant functions of carotenoids. *Free Radical Biology and Medicine*, 7(6), 617–635.

Kusmayadi, A., Leong, Y.K., Yen, H.-W., Huang, C.-Y., Chang, J.-S. (2021). Microalgae as sustainable food and feed sources for animals and humans – Biotechnological and environmental aspects. *Chemosphere*, 271, 129800.

Leonardi, P.I., Popovich, C.A., Damiani, M.C. (2011). Feedstocks for second-generation biodiesel: Microalgae's biology and oil composition. In *Economic Effects of Biofuel Production*, dos Santos Bernardes, M.A. (ed.). IntechOpen, 318–346.

Li, J., Stamato, M., Velliou, E. (2015). Design and characterization of a scalable airlift flat panel photobioreactor for microalgae cultivation. *J. Appl. Phycol.*, 27, 75–86.

Mansour, H., Shoman, S., Kdodier, M. (2011). Antiviral effect of edaphic cyanophytes on rabies and herpes-1 viruses. *Acta Biologica Hungarica*, 62(2), 194–203.

Marino, T., Casella, P., Sangiorgio, P., Verardi, A., Ferraro, A., Hristoforou, E., Musmarra, D. (2020). Natural beta-carotene: A microalgae derivate for nutraceutical applications. *Chemical Engineering Transactions*, 79, 103–108.

Marizcurrena, J.J., Cerdá, M.F., Alem, D., Castro-Sowinski, S. (2019). Living with pigments: The colour palette of Antarctic life. In *The Ecological Role of Microorganisms in the Antarctic Environment*, Castro-Sowinski, S. (ed.). Springer, Cham, 65–82.

Markets and Markets (2021). Carotenoid market by type (astaxanthin, beta-carotene, canthaxanthin, lutein, lycopene, and zeaxanthin), source (synthetic and natural), application (supplements, food, feed, and cosmetics), and by region: Global Forecast to 2027. Report [Online]. Available at: https://www.marketsandmarkets.com/Market-Reports/carotenoid-market-158421566.html.

Martău, G.A., Mihai, M., Vodnar, D.C. (2019). The use of chitosan, alginate, and pectin in the biomedical and food sector: Biocompatibility, bioadhesiveness, and biodegradability. *J. Polymers*, 1837, 11.

Mata, T.M., Martins, A.A., Caetano, N.S. (2010). Microalgae for biodiesel production and other applications: A review. *Renewable and Sustainable Energy Reviews*, 14(1), 217–232.

Mathews-Roth, M.M. and Krinsky, N.I. (1985). Carotenoid dose level and protection against UV-B induced skin tumors. *Photochemistry and Photobiology*, 42(1), 35–38.

Meléndez-Martínez, A.J., Britton, G., Vicario, I.M., Heredia, F.J. (2007). Relationship between the colour and the chemical structure of carotenoid pigments. *Food Chemistry*, 101(3), 1145–1150.

Minard, F. (2005). Method for the photostabilization of phycobiliproteins in an aqueous extract, compositions containing stabilised phycobiliproteins and use of phycobiliproteins. Patent, WO2005065697 Al.

Mishra, A., Kavita, K., Jha, B. (2011). Characterization of extracellular polymeric substances produced by micro-algae *Dunaliella salina*. *Carbohydr. Polym.*, 83, 852–857.

Nagarajan, D., Lee, D.J., Chang, J.-S. (2018). Heterotrophic microalgal cultivation. In *Bioreactors for Microbial Biomass and Energy Conversion*, Liao, Q., Chang, J.-S., Herrmann, C., Xia, A. (eds). Springer, New York.

Nalimova, A.A., Popova, V.V., Tsoglin, L.N., Pronina, N.A. (2005). The effect of copper and zinc on *Spirulina platensis* growth and heavy metal accumulation in its cell. *Russian Journal of Plant Physiology*, 52, 229–234.

Namikoshi, M. and Rinehart, K.L. (1996). Bioactive compounds produced by cyanobacteria. *Journal of Industrial Microbiology*, 17, 373–384.

Novoveská, L., Ross, M.E., Stanley, M.S., Pradelles, R., Wasiolek, V., Sassi, J.F. (2019). Microalgal carotenoids: A review of production, current markets, regulations, and future direction. *Marine Drugs*, 17(11), 640.

Ohlrogge, J. and Browse, J. (1995). Lipid biosynthesis. *The Plant Cell*, 7(7), 957.

Pagliaro, M. (2019). An industry in transition: The chemical industry and the megatrends driving its forthcoming transformation. *Angewandte Chemie International Edition*, 58(33), 11154–11159.

Patterson, G.M.L. (1996). Biotechnological applications of cyanobacteria. *J. Sci. Ind. Res.*, 55, 669–684.

Pereira, S.B., Sousa, A., Santos, M., Araújo, M., Serôdio, F., Granja, P., Tamagnini, P. (2019). Strategies to obtain designer polymers based on cyanobacterial extracellular polymeric substances (EPS). *International Journal of Molecular Sciences*, 20(22), 5693.

Perez-Garcia, O. and Bashan, Y. (2015). Microalgal heterotrophic and mixotrophic culturing for bio-refining: From metabolic routes to techno-economics. In *Algal Biorefineries. Volume 2: Products and Refinery Design*, Prokop, A., Bajpai, R., Zappi, M. (eds). Springer, Cham, 61–131.

Polaris Market Research (2022). Microalgae market share, size, trends, industry analysis report, by product type (Dunaliella Salina, Spirulina, Chlorella, others), microalgae strain (*Haematococcus Pluvialis*, *Phaeodactylum Tricornutum*, *Porphyridium Cruentum*, *Nannochloropsis*, others), grade (food grade, feed grade, fuel grade, cosmetics grade, others), application, by region, and segment forecast. Report [Online]. Available at: https://www.polarismarketresearch.com/industry-analysis/microalgae-market.

Puyfoulhoux, G., Rouanet, J.M., Besançon, P., Baroux, B., Baccou, J.C., Caporiccio, B. (2001). Iron availability from iron-fortified spirulina by an in vitro digestion/Caco-2 cell culture model. *Journal of Agricultural and Food Chemistry*, 49(3), 1625–1629.

Raposo, M.F.J., De Morais, A.M.M.B., De Morais, R.M.S.C. (2014). Influence of sulphate on the composition and antibacterial and antiviral properties of the exopolysaccharide from *Porphyridium cruentum*. *Life Sciences*, 101, 56–63.

Rastogi, R.P. and Incharoensakdi A. (2014). Characterization of UV-screening compounds, mycosporine-like amino acids, and scytonemin in the cyanobacterium *Lyngbya sp.* CU2555. *FEMS Microbiol. Ecol.*, 87, 244–256.

Rastogi, R.P., Sonani, R.R., Madamwar, D. (2015). Cyanobacterial sunscreen scytonemin: Role in photoprotection and biomedical research. *Appl. Biochem. Biotechnol.*, 176(6), 1551–1563.

Rastoin, J.-L. (2017). Le secteur des micro-algues en Méditerranée. Perspectives et contribution au développement durable. Report, IPEMED.

Remize, M., Brunel, Y., Silva, J.L., Berthon, J.Y., Filaire, E. (2021). Microalgae n-3 PUFAs production and use in food and feed industries. *Marine Drugs*, 19(2), 113.

Research and Markets (2021a). Microalgae market by type (Spirulina, Chlorella, Dunaliella Salina, and Aphanizomenon Flos-Aquae), application (dietary supplements, food/feed, pharmaceutical, cosmetic, and others): Global opportunity analysis and industry forecast 2021-2028. Report [Online]. Available at: https://www.alliedmarketresearch.com/microalgae-market-A13419.

Research and Markets (2021b). Phycocyanin market by form (powder and liquid), grade (food grade, pharma grade, and reagent and analytical grade) and application (food and beverages, pharmaceuticals, cosmetics, nutraceuticals, and others): Global opportunity analysis and industry. Forecast Report [Online]. Available at: https://www.alliedmarketresearch.com/phycocyanin-market-A12344.

Rothschild, L.J. and Mancinelli, R.L. (2001). Life in extreme environments. *Nature*, 409(6823), 1092–1101.

Schneiderman, D.K. and Hillmyer, M.A. (2017). 50th anniversary perspective: There is a great future in sustainable polymers. *Macromolecules*, 50(10), 3733–3749.

Schoenfeld, D. (2005). Fatty acid antiarrhythmia trial investigators. *Circulation*, 112, 2762–2768.

Schümann, K., Ettle, T., Szegner, B., Elsenhans, B., Solomons, N.W. (2007). On risks and benefits of iron supplementation recommendations for iron intake revisited. *J. Trace Elem. Med. Biol.*, 21, 147–168.

Sharma, J., Sarmah, P., Bishnoi, N.R. (2020). Market perspective of EPA and DHA production from microalgae. In *Nutraceutical Fatty Acids from Oleaginous Microalgae: A Human Health Perspective*, Patel A.K., Matsakas L. (eds). Wiley Online Library, Hoboken.

Show, P.L. (2022). Global market and economic analysis of microalgae technology: Status and perspectives. *Bioresource Technology*, 357, 127329.

Sinha, R.P. and Häder, D.P. (2008). UV-protectants in cyanobacteria. *Plant Science*, 174(3), 278–289.

Skjånes, K., Aesoy, R., Herfindal, L., Skomedal, H. (2021). Bioactive peptides from microalgae: Focus on anti-cancer and immunomodulating activity. *Physiologia Plantarum*, 173(2), 612–623.

Soanen, N., Da Silva, E., Gardarin, C., Michaud, P., Laroche, C. (2016). Improvement of exopolysaccharide production by *Porphyridium marinum. Bioresour. Technol.*, 213, 231–238.

Soto-Sánchez, O., Hidalgo, P., González, A., Oliveira, P.E., Hernández Arias, A.J., Dantagnan, P. (2023). Microalgae as raw materials for aquafeeds: Growth kinetics and improvement strategies of polyunsaturated fatty acids production. *Aquaculture Nutrition*, 5110281.

Spolaore, P., Joannis-Cassan, C., Duran, E., Isambert, A. (2006). Commercial applications of microalgae. *Journal of Bioscience and Bioengineering*, 101(2), 87–96.

Steiner, B.M., McClements, D.J., Davidov-Pardo, G. (2018). Encapsulation systems for lutein: A review. *Trends in Food Science and Technology*, 82, 71–81.

Stevenson, C.S., Capper, E.A., Roshak, A.K., Marquez, B., Eichman, C., Jackson, J.R., Mattern, M., Gerwick, W.H., Jacobs, R.S., Marshall, L.A. (2002). The identification and characterization of the marine natural product scytonemin as a novel antiproliferative pharmacophore. *Journal of Pharmacology and Experimental Therapeutics*, 303, 858–866.

Transparency Market Research (2019). Product: Spirulina, Chlorella, astaxanthin, beta carotene, PUFA fatty acid, lutein, and others; application: Food and beverages, personal care, nutraceuticals and dietary supplements, pharmaceuticals, chemicals, fuel, animal feed, and others. Report, ASDR-505008.

Verdelho Viera, V. (2015). Overview of production of value added products. Markets and producers in the EU. Report, Pullach, Munich.

Volk, R.B. and Furkert, F.H. (2006). Antialgal, antibacterial and antifungal activity of two metabolites produced and excreted by cyanobacteria during growth. *Microbiological Research*, 161(2), 180–186.

Vu, B., Chen, M., Crawford, R.J., Ivanova, E.P. (2009). Bacterial extracellular polysaccharides involved in biofilm formation. *Molecules*, 14(7), 2535–2554.

Wada, N., Sakamoto, T., Matsugo, S. (2015). Mycosporine-like amino acids and their derivatives as natural antioxidants. *Antioxidants*, 4, 603–646.

Wang, Y., Tibbetts, S.M., McGinn, P.J. (2021). Microalgae as sources of high-quality protein for human food and protein supplements. *Foods*, 10(12), 3002.

Wicaksono, H.A., Satyantini, W.H., Masithah, E.D. (2019). The spectrum of light and nutrients required to increase the production of phycocyanin *Spirulina platensis*. *IOP Conference Series: Earth and Environmental Science*, 236(1), 012008.

Winwood, R.J. (2013). Recent developments in the commercial production of DHA and EPA rich oils from micro-algae. *OCL*, 20(6), D604.

Xiang, G., Xin, J., Xufeng, L., Lindblad, P. (2021). Biotechnological production of the sunscreen pigment scytonemin in cyanobacteria: Progress and strategy. *Mar. Drugs*, 19(3), 129.

Xiao, R. and Zheng, Y. (2016). Overview of microalgal extracellular polymeric substances (EPS) and their applications. *Biotechnology Advances*, 34(7), 1225–1244.

Yashodhara, B.M., Umakanth, S., Pappachan, J.M., Bhat, S.K., Kamath, R., Choo, B.H. (2009). Omega-3 fatty acids: A comprehensive review of their role in health and disease. *Postgraduate Medical Journal*, 85(1000), 84–90.

Zaccaro, M.C., Kato, A., Zulpa, G., Storni, M.M., Steyerthal, N., Lobasso, K., Stella, A.M. (2006). Bioactivity of *Scytonema hofmanni* (Cyanobacteria) in *Lilium alexandrae* in vitro propagation. *Electronic Journal of Biotechnology*, 9(3), 210–214.

Zheng, W., Caifa, C., Qiping, C., Yiqin, A., Chengcai, C. (2006). Oral administration of exopolysaccharide from *Aphanothece halophytica* (Chroococcales) significantly inhibits influenza virus (H1N1)-induced pneumonia in mice. *Int. Immunopharmacol.*, 6, 1093–1099.

Zheng, H., Wang, Y., Li, S., Nagarajan, D., Varjani, S., Lee, D.J., Chang, J.S. (2022). Recent advances in lutein production from microalgae. *Renewable and Sustainable Energy Reviews*, 153, 111795.

Zia-Ul-Haq, M. (2021). Historical and introductory aspects of carotenoids. In *Carotenoids: Structure and Function in the Human Body*, Zia-Ul-Haq, M., Dewanjee, S., Riaz, M. (eds). Springer, Berlin/Heidelberg, 1–42.

Zili, F., Mezhoud, N., Trabelsi, L., Chreif, I., Ben Ouada, H. (2015). Fatty acid composition of the thermophilic *Gloeocapsa gelatinosa* under different combinations of temperature, light intensity, and $NaNO_3$ concentration. *Journal of Applied Phycology*, 27(1), 97–107.

8

Economic and Environmental Optimization of Fermentation Processes: Ethanol and Glutamic Acid Production

Jean-Marc ENGASSER and Mohamed GHOUL

Laboratoire Réaction et Génie des Procédés (LRGP), CNRS, Université de Lorraine, Nancy, France

Optimizing the use of microbial fermentation or microalgae cultivation to produce cells or metabolites involves finding the operating mode and variables that will reduce production costs. However, this optimization cannot be limited to the fermentation stage alone, as the concentrations of cells and metabolites reached during fermentation can influence the sizing of other operations in the process. Consequently, a global approach to optimization must be adopted, taking into account all of the upstream and downstream operations that make up the process.

The economic approach aimed at reducing production costs can be complemented by an environmental optimization approach, aimed at reducing the environmental impacts generated by the production process. This chapter presents a brief overview of the methodology used to assess the production cost and environmental indicators of fermentation processes and products. We then examine its application to the economic and environmental optimization of two fermentations for the production of ethanol and glutamic acid.

Bioreactor Implementation in the Agro-Food Industries,
coordinated by Mohamed GHOUL.

8.1. Economic assessment and optimization of fermentation processes

8.1.1. *Methodology for economic assessment of fermentation processes*

The economic assessment of a fermentation process is based on the sizing of the various unit operations in the process, including their material balances, energy requirements and equipment sizes (Figure 8.1) (Chauvel et al. 2003; Peters et al. 2003; Silla 2003; Towler and Sinnott 2021).

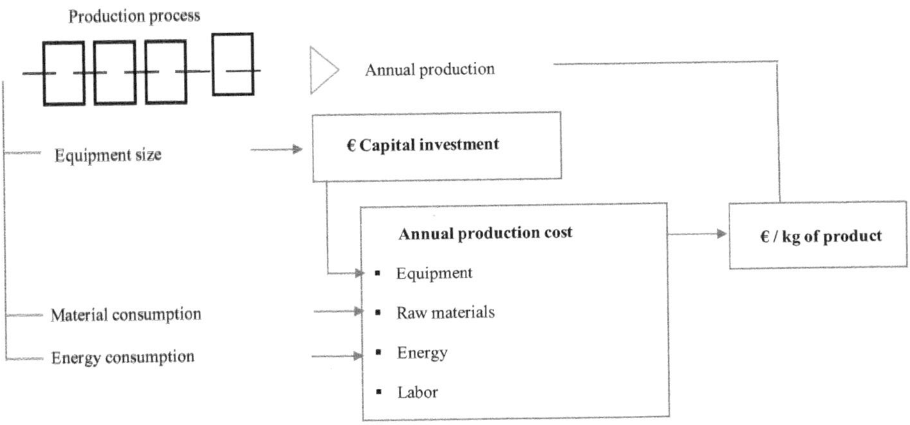

Figure 8.1. *Methodology for the economic assessment of processes*

The first step in this assessment is to estimate the initial investment required, also known as CAPEX (capital expenditure). This initial capital investment takes into account the cost of acquiring the equipment, which depends on its size, as well as the additional costs required to build the production unit, install the equipment and commission it.

The second step is to determine the cost of production per year or per kilogram of product, also known as OPEX (operational expenditure). This assessment encompasses various elements such as equipment, raw materials, energy and labor costs.

In terms of equipment operating costs, the first step is to calculate the annual depreciation of the investment in equipment, as well as the associated operating costs. In addition, any additional costs associated with the use of equipment consumables such as membranes, filters or resins must be considered. Based on the process material balance, the annual cost of raw materials, reagents and water used, and the annual costs associated with effluent and waste treatment are calculated by multiplying the quantities of materials involved by their unit costs. Similarly, the annual cost of energy is determined by multiplying the consumption of energy vectors (fuel oil, gas and electricity) by their unit costs. Lastly, the cost of labor is assessed on the basis of staffing requirements to operate the production unit and employee salary costs.

To obtain the total annual production cost, all of these contributions are added together. Then, by dividing the annual operating cost by the annual quantity of product manufactured, we obtain the unit production cost per kilogram or ton of product.

8.1.2. *Range of cost of fermentation processes*

The capital and production costs of fermentation processes vary considerably and are strongly influenced by the nature and purity of the product, as well as the scale of production. As a general rule, investment costs can vary from tens of millions of euros to several hundreds of millions of euros.

Production costs are just as wide-ranging, from pennies to millions of euros per kilogram of product. The lowest costs, below one cent per kilogram, are those relating to biological water and effluent treatment. Large-scale bioproducts, such as ethanol, organic and amino acids or polysaccharides, are produced at costs of between 1 and 10 euros per kilogram. For specialty chemicals, as well as drugs such as antibiotics, production costs generally range from tens to hundreds of euros per kilogram. Lastly, the highest production costs are for therapeutic proteins and vaccines, generally reaching hundreds of thousands, even millions of euros per kilogram.

8.1.3. *Process cost analysis*

In the context of an economic optimization of a fermentation process, it is a good idea to take a closer look at production costs by examining their origins. An appropriate analysis involves identifying the production cost ratios attributed to raw materials, energy consumption, equipment use and labor expenses. It is also worth determining the contribution of the upstream sector of the process, covering media preparation and bioprocessing steps, as well as the downstream sector, which includes separation and purification operations.

This analytical approach makes it possible to pinpoint areas where optimizations can be implemented to reduce production costs. In terms of upstream operations, costs can be reduced by increasing fermenter productivity, optimizing substrate-to-product conversion efficiency or selecting less expensive raw materials. In downstream operations, cost reduction can include improving product separation efficiency, reducing energy consumption, and automating operations to reduce labor requirements.

8.2. Environmental assessment and optimization of fermentation processes

8.2.1. *Environmental impacts of lifecycle assessment*

The environmental assessment of a process can be carried out using the lifecycle assessment (LCA) methodology. LCA represents a comprehensive approach that aims to assess the environmental footprint of a product or process throughout its lifecycle (Figure 8.2) (Klöpffer and Grahl 2014; Shaked et al. 2015; Hauschild et al. 2018).

LCA takes various types of environmental impact into account. First, it assesses impacts linked to resource consumption, including primary energy consumption (PEC), depletion of abiotic resources and water consumption (WAT). Second, it analyzes the impacts resulting from emissions into the air, water or soil, such as the global warming potential (GWP) of greenhouse gases, the formation of photochemical oxidants or the eutrophication of water. LCA also takes into account global toxicity impacts, such as the potential for human toxicity.

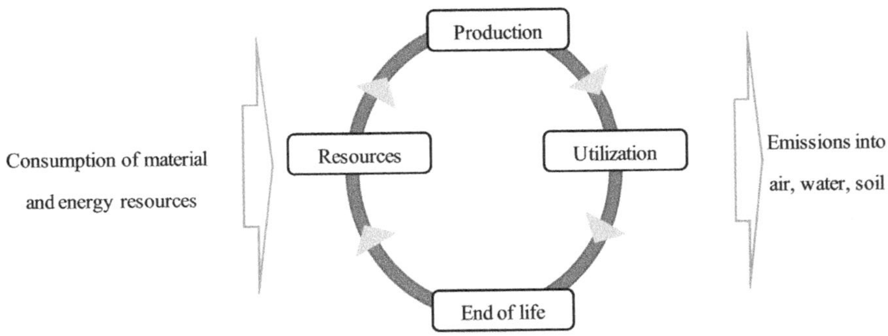

Figure 8.2. *Lifecycle assessment of processes and products*

When LCA is applied to a product, it examines all stages of its lifecycle, from the production of raw materials to the manufacture, use and end-of-life of the product. LCA can also be applied more narrowly to a product's production process, which is particularly useful for comparing and optimizing processes. In this context, LCA considers the environmental impacts generated by the unit operations of the process, as well as the upstream impacts generated by the production and transport of material and energy inputs. For the environmental assessment of a fermentation process, the most important environmental impacts to consider are PEC, WAT and GWP.

8.2.2. *Assessment of environmental impact indicators*

In the lifecycle assessment method, environmental impacts are assessed using specific indicators (Figure 8.3). These impact indicators are quantities that quantify the environmental contributions of all material and energy inputs and outputs. Each of these impact indicators has its own unit of measurement. For example, PEC is expressed in Megajoules, WAT in kilograms of water, and GWP in equivalent kilograms of CO_2.

An impact indicator (I_{IMP}) is calculated by summing up all material and energy inputs and outputs, according to the following relationship:

$$I_{IMP} = \sum (m_i * F_i) \qquad [8.1]$$

In this equation, m_i represents input and output quantities determined from inventories or material and energy balances, while F_i designates impact factors per unit of input or output quantity. For a material input, the impact factor is the impact indicator per kilogram, assessed on the basis of an LCA study of the input. It takes into account all amounts that impact during the input's extraction, production and transport stages. Impact factor values are available in environmental databases. For an energy vector, such as fuel, electricity or gas, the impact factor per unit quantity is also assessed on the basis of an LCA study of the carrier. This LCA considers all environmental contributions throughout the vector's lifecycle, covering the stages of extraction, production, transport and use. For example, for gas, the impact factor includes amounts from components emitted during gas combustion. For CO_2, its impact factor in terms of GWP is set at 1, since CO_2 is the reference compound for assessing GWP impact.

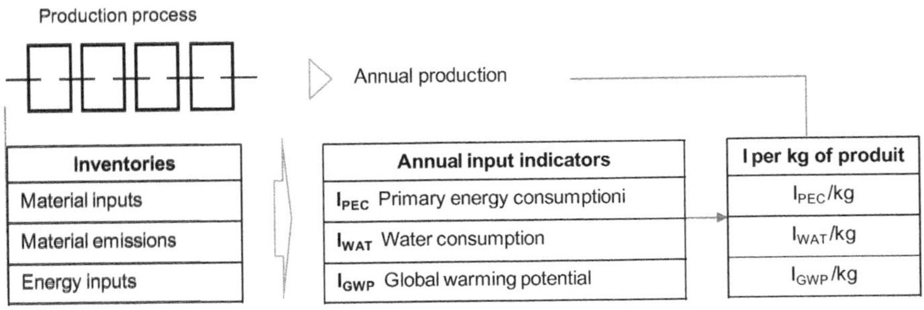

Figure 8.3. *Methodology for assessing environmental impact indicators for processes*

Using this LCA methodology, the impact indicators for a fermentation process not only include the impacts generated during the various unit operations of the process, but also those associated with obtaining and transporting raw materials and energy vectors. As a result, the primary energy and WATs, as well as the CO_2 emissions assessed by these indicators, are higher than the consumptions and emissions assessed on the basis of the process' material and energy balances alone.

8.2.3. *Environmental analysis and optimization of fermentation processes*

It is a good idea to supplement the assessment of environmental impact indicators with an analysis of their origins. This approach makes it possible, for example, to specify the proportions of impact indicators that can be attributed to the use of raw materials, energy consumption and pollutant emissions. It also makes it possible to quantify the relative contribution of the various unit operations of the process to these impact indicators.

Based on this analytical approach, it becomes possible to identify the areas in which optimizations can be implemented, so as to reduce the environmental impact of a process. As far as fermentation processes are concerned, the main eco-design approaches to be favored are (i) reducing raw material consumption, in particular WAT, which can be achieved, for example, by recycling water between different unit operations; (ii) reducing energy consumption, which is often achieved by thermal integration or heat recycling between operations; and (iii) reducing pollutant and effluent emissions during process operations.

By implementing these approaches, it is possible to significantly improve the environmental footprint of fermentation processes.

8.3. Optimization of the ethanol fermentation process

8.3.1. *Ethanol production processes*

Bioethanol, as a fuel, is the most widely produced compound by microbial fermentation. By 2022, global production had reached around 90 million tons, with the United States and Brazil leading the way. In Europe, bioethanol production stands at around 4 million tons. Ethanol is obtained from various agricultural raw materials, including sugarcane, sugar beet, corn, wheat and barley (Cardona et al. 2007; Geddes et al. 2011).

The yeast *Saccharomyces cerevisiae* is widely used for its ability to convert glucose and sucrose into ethanol, and for its tolerance of high ethanol concentrations. Recombinant strains of *S. cerevisiae* have also been developed to ferment xylose present in hemicellulose hydrolysates.

In addition to the fermentation stage, industrial production processes involve a series of unitary mechanical, thermal and separation operations (Figure 8.4). Upstream of fermentation, these operations aim to convert plant raw materials into a sugar solution for fermentation. For example, sugarcane stalks are crushed to extract the sweet juice using mechanical rollers or hydraulic presses. Sugar beet undergoes a washing process followed by cutting into cossettes, and then undergoes sugar extraction using hot water. Corn kernels are initially ground, then subjected to high-temperature liquefaction and enzymatic hydrolysis using amylases and glucoamylases in order to convert starch into glucose. For lignocellulosic raw materials, pretreatment techniques include mechanical grinding, acid hydrolysis of hemicellulose to xylose and enzymatic hydrolysis of cellulose to glucose using cellulases.

After the fermentation stage, the fermented medium containing ethanol, residual sugars, yeast cells and other excreted metabolites, is subjected to seperation operations to obtain purified ethanol. Distillation is the most widely used method for concentrating and purifying ethanol. If the aim is to obtain anhydrous ethanol, an additional step is molecular sieve dehydration or azeotropic distillation.

Ethanol production processes generate large quantities of by-products, both during the initial sugar extraction phase and during the final ethanol recovery phase.

The quantity of these by-products often exceeds the quantity of ethanol produced. Additional operations are carried out to process and add value to these by-products. For example, after the sugar has been extracted from beet, the residual pulp is pressed and dried for use as animal feed. In the case of sugar cane, the fibrous residue known as bagasse is generally incinerated to produce heat and electricity for internal use. As for distillation vinasses, they are concentrated by evaporation and used as fertilizers in agricultural practices. When corn starch is transformed into ethanol, the remaining solid residues are separated, dried and transformed into distiller's dried grains, which are then used for animal feed.

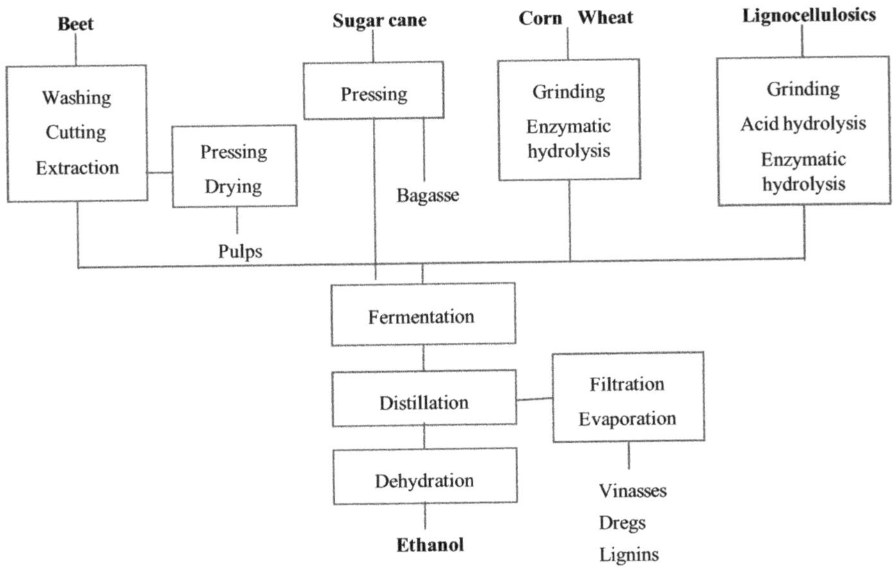

Figure 8.4. *Processes for producing ethanol from agro-resources*

8.3.2. *Kinetics of ethanol fermentation*

The kinetics of anaerobic sugar transformation by the yeast *Saccharomyces cerevisiae* can be understood in a relatively simple way, as ethanol production takes place throughout microbial growth. Fermentation generally takes place in a temperature range from 25 to 35°C, with pH values between 4.0 and 6.0. Optimum conditions vary according to the yeast strain used and the composition of the fermentation medium.

In order to have a fermentation simulation model, it is necessary to postulate the kinetic laws that govern the rates of yeast growth, ethanol production, as well as sugar consumption (Ariyajaroenwong et al. 2016). The specific growth rate of yeast is limited at low sugar concentrations and is inhibited at high sugar concentrations. Furthermore, it also undergoes inhibition at high ethanol levels. The law proposed to describe the relationship between the specific growth rate μ and the concentrations of sugar S and ethanol Eth is as follows:

$$\mu = \left(\frac{\mu_{max}\, S}{K_S + S + \frac{S^2}{K_{Si}}} \right) \left(1 - \frac{Eth}{Eth_m}\right)^{\alpha} \qquad [8.2]$$

Since ethanol production is associated with growth, its rate is also limited at low sugar concentrations and inhibited at high sugar and ethanol concentrations. The law describing the specific ethanol production rate q_{Eth} as a function of sugar and ethanol concentrations is as follows:

$$q_{Eth} = \left(\frac{q_{Eth,max}\, S}{K_{SE} + S + \frac{S^2}{K_{SiE}}} \right) \left(1 - \frac{Eth}{Eth_{mE}}\right)^{\beta} \qquad [8.3]$$

It should be noted that the inhibition constants for ethanol production generally differ from those for the growth law. For example, the maximum ethanol concentration for ethanol production, Eth_{mE}, is generally higher than that for growth, Eth_m, meaning that ethanol production can persist even after total growth inhibition by ethanol.

The kinetic law for the specific sugar consumption rate q_S comprises three terms:

$$q_S = Y_{S/X}\, \mu + Y_{S/Eth}\, q_{Eth} + m_S \qquad [8.4]$$

The first term is sugar consumption proportional to the specific rate of growth, the second is sugar consumption proportional to the specific rate of ethanol production, and the third is the rate of sugar consumption for maintenance.

The kinetic parameters for growth, ethanol production and sugar consumption depend on the yeast strain, temperature and pH of the medium. In the simulations carried out to optimize ethanol production, the kinetic parameters used are shown in Table 8.1.

Yeast growth	Ethanol production	Sugar consumption
μ_{max}: 0.3 h^{-1} K_S: 20 g/L K_{SI}: 290 g/L Eth_m: 110 g/L α: 1.5	$q_{Eth,max}$: 3 gEth/gX.h K_{SE}: 20 g/L K_{SIE}: 350 g/L Eth_{mE}: 125 g/L β: 1.5	$Y_{S/X}$: 2 gS/gX $Y_{S/Eth}$: 2.2 gS/gEth ms: 0.001 h^{-1}

Table 8.1. *Kinetic parameters of the ethanol fermentation simulation model*

8.3.3. *Optimum implementation of ethanol fermentation*

8.3.3.1. *Economic optimization of ethanol production*

The aim of optimizing the ethanol production process is to reduce the cost of producing this biofuel as much as possible. This is essential, because bioethanol, derived from plant raw materials, must be economically competitive with fossil fuels.

For the fermentation operation, cost minimization involves two key aspects: first, increasing fermenter productivity to reduce equipment, energy and labor costs per kilogram of ethanol produced, and second, improving overall sugar-to-ethanol conversion efficiency to reduce the amount of raw materials. When considering the subsequent ethanol distillation stage, another factor influencing production costs is the ethanol concentration obtained at the end of fermentation. A lower ethanol concentration means higher energy consumption for ethanol concentration. Consequently, optimization of the fermentation process must take into account the costs associated with the entire production process.

Economic optimization of the process can be driven by knowledge of the distribution of costs according to their origins. At the industrial production scale, cost breakdown analysis shows that the carbon substrate accounts for around half of the total cost of ethanol production, while energy contributes to around 25%. Equipment and labor costs, meanwhile, account for between 10% and 15% of overall production costs. This means that to optimize the ethanol production process economically, it is imperative to maximize the efficiency of sugar utilization during fermentation and to reduce energy consumption in other operations.

8.3.3.2. *Comparison of batch and continuous fermentations*

As explained in the chapter on fermentations, anaerobic fermentation of a metabolite such as ethanol can either be carried out in a batch or continuous mode.

However, for a growth-related metabolite, fed-batch fermentation is not competitive due to the reduced productivity, resulting from partial utilization of the fermenter volume. In this section, we carry out a comparison of optimal operating conditions between batch and continuous fermentations, with the aim of determining the lowest production cost. This comparison is based on the annual utilization of a 1,000 cubic meter fermenter.

When fermentation is carried out in batch mode, the main operational variable to be optimized is the initial sugar concentration. At a given sugar concentration, the fermentation time is adjusted to convert at least 99% of the sugar. For the yeast under consideration, the simulation of the batch fermentation and the assessment of the cost of ethanol production indicate that a minimum production cost is achieved when the initial sugar concentration is around 200 g/L. Under these conditions, fermentation lasts around 40 h. Time-dependent variations in sugar, biomass and ethanol concentrations are shown in Figure 8.5(a). Final yeast and ethanol concentrations, overall sugar conversion efficiency, fermenter productivity, annual production and cost are summarized in Table 8.2. For batch fermentation, the optimal production cost is around 1.40 € per kg of ethanol. The main contributions of this cost come from raw materials, mainly sugar, accounting for over 60% of the production cost, and energy, which contributes around 20% (Figure 8.6).

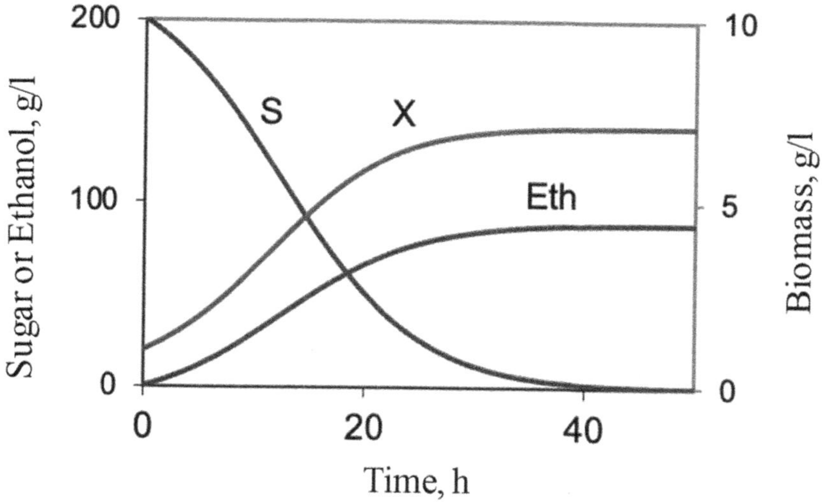

Figure 8.5. *(a) Variation in sugar S, biomass X and Ethanol Eth concentrations during optimized batch fermentation*

	X, g/L	Eth, g/L	$Y_{Eth/S\ glob}$, gEth/sS	Prod $_{Eth}$, gEth/L·h	Cost, €/kg	Annual Eth, tEth
Batch	7	90	0.44	1.8	1.4	15,000
Continuous	5	70	0.40	1.4	1.6	11,000

Table 8.2. *Technical and economic characteristics of optimized ethanol fermentation in batch and continuous mode*

When fermentation is run in continuous mode, the main operational variable to be optimized is the sugar concentration in the feed. The initial concentration is identical to the feed concentration. At a given sugar concentration, the dilution rate or residence time of the medium in the fermenter is adjusted to convert the maximum amount of sugar. Simulation of continuous fermentation of sugar into ethanol indicates a minimum cost of ethanol production at a sugar feed concentration of around 170 g/L and a residence time of around 50 h. Under these optimum conditions, the variations over time in sugar, biomass and ethanol concentrations are shown in Figure 8.5(b). After the initial batch fermentation, a steady state is reached after around 100 h. Stationary yeast and ethanol concentrations, overall sugar conversion efficiency, fermenter productivity, annual production and ethanol production cost are shown in Table 8.2, with corresponding additions to production cost in Figure 8.6.

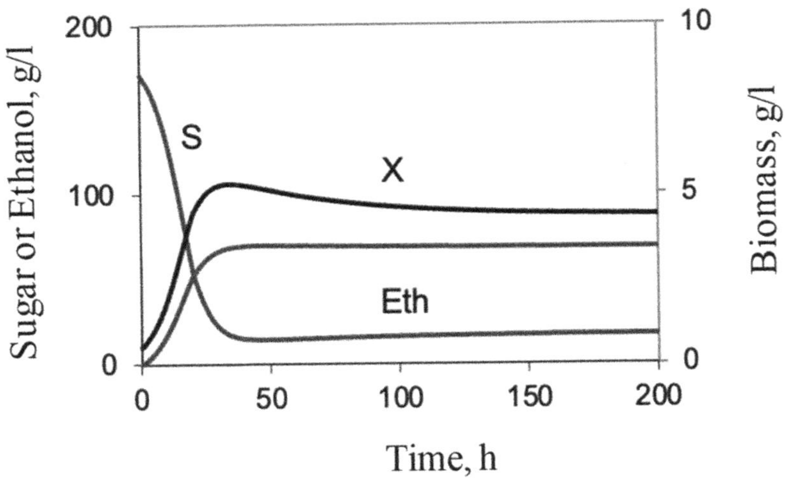

Figure 8.5. *(b) Variation in sugar S, biomass X and Ethanol Eth concentrations during optimized continuous fermentation*

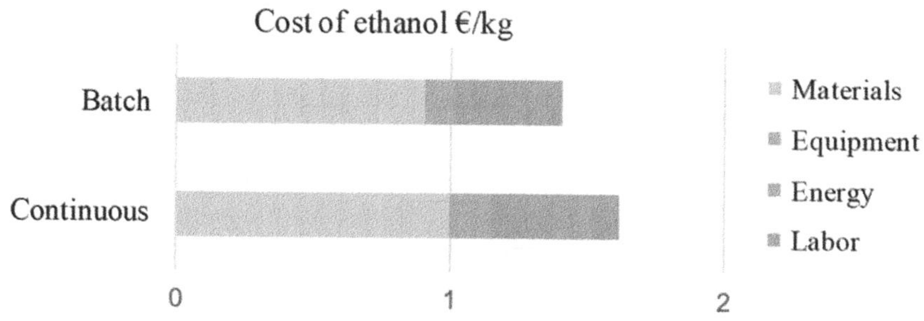

Figure 8.6. *Batch and continuous fermenter ethanol production cost analysis by origin. For a color version of this figure, see www.iste.co.uk/ghoul/bioreactor.zip*

A comparison of the performance of the two fermentation modes shows that batch fermentation offers a clear advantage for ethanol production. Batch fermentation results in higher ethanol concentration, higher sugar conversion efficiency and higher fermenter productivity than continuous fermentation. Despite significantly reduced downtimes in the continuous fermentation process, the reduced productivity stems from the significant inhibition of ethanol production by ethanol. In the case of batch fermentation, inhibition increases progressively as the concentration of ethanol in the medium increases. In continuous mode, once the steady state has been reached, the entire medium in the fermenter is at the highest ethanol concentration, meaning that the cells are constantly subjected to maximum inhibition. This reduced productivity in continuous fermentation leads to a sharp drop in annual ethanol production and a significant increase in production costs. As a result, batch fermentation is proving to be the most technically and economically competitive mode of operation.

8.3.4. *Possibilities for intensifying ethanol fermentation*

Various technological alternatives can be considered to improve the performance of continuous ethanol fermentation (Bayrock and Michael Ingledew 2001; Puligundla et al. 2019). The first approach is to adopt a serial fermentation system, where several fermenters are used in succession instead of a single fermenter (Figure 8.7). In this cascade scheme, the ethanol concentration increases progressively from one fermenter to the next, thereby reducing ethanol inhibition of yeast. This means that only yeasts in the last fermenter are exposed to the highest ethanol concentration. In addition, it is possible to lower the fermentation temperature at each stage to increase yeast tolerance to high ethanol concentrations.

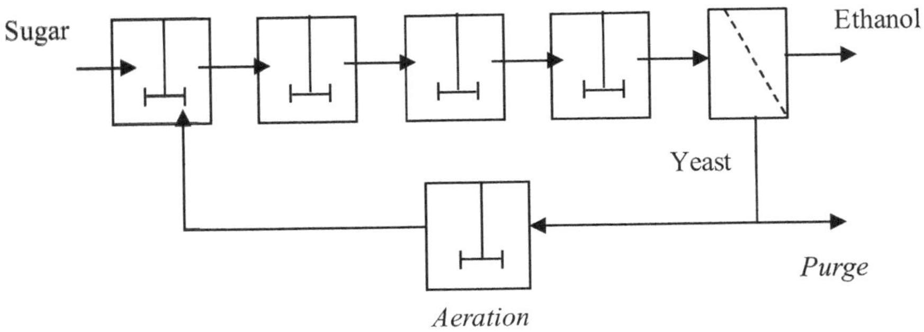

Figure 8.7. *Multi-stage continuous fermenter with cell recycling for ethanol production*

A second, additional approach is to recycle and regenerate the yeast. At the outlet of the last stage, the yeasts are recovered using methods such as centrifugation or microfiltration, then reintroduced into the first stage of the cascade. During this recycling process, yeast can also be regenerated by aeration. In order to maintain a stable yeast concentration in this operating mode, continuous cell removal can be integrated into the yeast recycling circuit.

The intensification of ethanol production in this configuration of cascade fermenters can be predicted by simulation. For example, for a cascade comprising four fermenters, with 15% recycle and yeast purge, simulation results are shown in Table 8.3. Compared with single-stage continuous fermentation, the multi-stage configuration leads to a much higher steady yeast concentration (35 g/L instead of 5 g/L), considerably higher fermenter productivity (multiplied by a factor of 20) and significantly reduced residence times. These productivity gain predictions are corroborated by experimental results obtained on multi-stage fermenters at the laboratory and pilot scale.

	S_0 g/l	τ, h	X, g/L	Eth, g/L	$Y_{Eth/S\ glob}$, gEth/gS	Prod Eth, gEth/L.h
Multi-stage	200	15	34	90	0.44	16

Table 8.3. *Technical characteristics of ethanol production in a multi-stage continuous fermenter with yeast recycling*

Industrial implementation of this fermenter technology requires a number of factors to be considered. Compared to batch fermentation, the implementation of continuous multi-stage fermentation with cell recycling is more complex. It is also more vulnerable to contamination by lactic acid bacteria. The production of organic acids by these bacteria leads to acidification of the medium, reducing the yield of sugar converted into ethanol. From an economic point of view, productivity gains translate into reductions in equipment and labor costs per kilogram of ethanol. However, raw material and energy costs remain comparable to those of batch fermentation. Further economic analysis is required to clarify the potential economic benefits of multi-stage technology.

8.3.5. *Environmental optimization of the ethanol production process*

Alongside the search for the lowest production cost, the optimization of the ethanol production process also aims to reduce its environmental impact. For ethanol-based biofuel obtained from renewable resources, it is essential to obtain a more favorable environmental footprint than that of a fossil fuel, such as gasoline. The upstream and downstream operations of the ethanol production process are characterized by high thermal energy consumption for the extraction, distillation, evaporation and drying stages, high WAT for extraction and significant CO_2 emissions during fermentation. Consequently, the three most significant LCA impacts of the ethanol production process are PEC, WAT and GWP.

Operating the fermenter to achieve the highest productivity and sugar conversion efficiency, which is essential for minimizing production costs, also helps to reduce environmental impact indicators per kilogram of ethanol. Several additional approaches can also be implemented in the production process to reduce the environmental impact of bioethanol. These include recovering the large quantities of CO_2 emitted during fermentation (around 0.45 g of CO_2 per g of sugar used), saving water, mainly by recycling water, and saving energy, mainly by recycling heat.

The potential environmental benefits of process optimization are illustrated using the example of ethanol production from sugar beet (Figure 8.8). The modifications made to the production process are as follows: first, the recovery of the CO_2 emitted during the fermentation stage by cleaning and compressing the gas phase leaving the fermenter; second, the recycling of the water used to wash the beets; third, the use of the water recovered during the evaporation stage for the sugar extraction operation; and lastly, the recycling of heat or thermal integration. For the distillation operation, the heat recovered during condensation of the vapor phase can be used to partially preheat the feed medium. Heat can also be recovered from the hot condensate and vapor phases of evaporation, and transferred to distillation, extraction and air-drying operations. Assessment of the impact of these modifications on the three bioethanol impact indicators for the entire bioethanol life cycle shows a reduction of around

40% in PEC, 50% in WAT and 60% in GWP. Under these optimized conditions, bioethanol's footprint is characterized by the three impact indicator values per kilogram of ethanol produced: 22 MJ/kg for PEC, 11 kg water/kg for WAT and 0.6 kg CO_2/kg for GWP. Compared with gasoline impact indicators, bioethanol has higher WAT, but significantly lower PEC and CO_2 production.

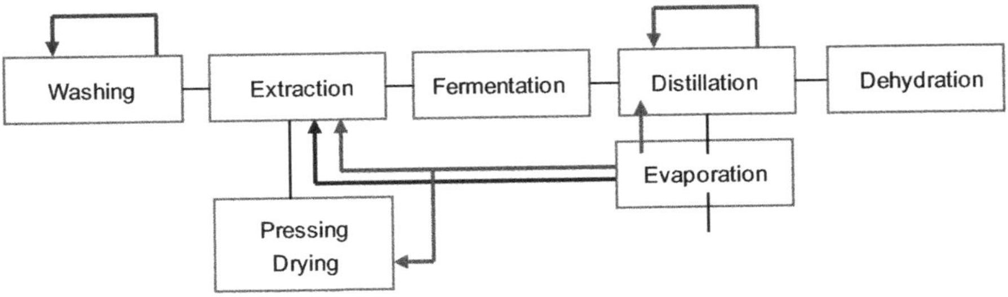

Figure 8.8. *Water and heat recycling strategies for the ethanol production process. For a color version of this figure, see www.iste.co.uk/ghoul/bioreactor.zip*

8.4. Optimization of the glutamic acid fermentation process

8.4.1. *Glutamic acid production process*

Glutamic acid is one of the leading industrial amino acids, with annual production reaching 3 million tons. The main raw materials used are beet or cane sugar and starch hydrolysates. Monosodium glutamate, the sodium salt of glutamic acid, is widely used in the food industry, mainly as a flavor enhancer. It should be noted that glutamic acid is generally referred to as glutamate, as it mainly occurs as negatively charged glutamate at neutral pH.

Glutamic acid is produced industrially by strains of *Corynebacterium glutamicum*. In these corynebacteria, as in many other microbial cells, glutamic acid is one of the metabolites with the highest intracellular concentration. It is produced by the cells from the ammonia they consume, before being converted into other amino acids. Depending on the strain of *C. glutamicum*, several techniques are used to induce glutamate excretion. These include limiting biotin in the fermentation medium, adding surfactants or antibiotics, and raising the temperature of the medium (Kumar et al. 2014; Wendisch et al. 2016).

The industrial glutamic acid production process (Figure 8.9) includes upstream operations for medium preparation, sterilization and fermentation. Then, downstream operations include centrifugation of the fermented medium to remove bacterial cells, evaporation to concentrate the glutamate, crystallization and, lastly, drying of the glutamate crystals. It can also include an additional step of acid hydrolysis of the cell biomass in order to recover the glutamate contained in the concentrated biomass phase. The by-products of the process are the residual vinasses from glutamate crystallization. This vinasse is concentrated by evaporation and used as animal feed.

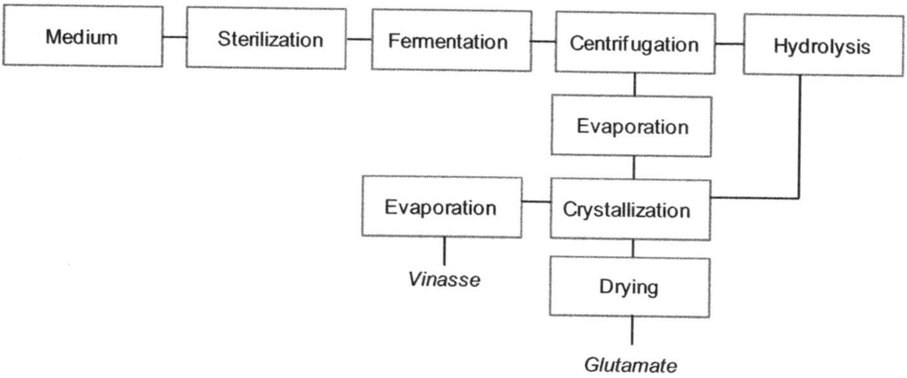

Figure 8.9. *Industrial glutamic acid production process*

8.4.2. *Kinetics of glutamic acid fermentation*

From the perspective of fermentation kinetics, glutamate behaves as a metabolite not associated with cell growth (Zhang et al. 1998; Delaunay et al. 1999). Fermentation begins with an initial phase of rapid growth, during which cells consume sugar without excreting metabolites. Once glutamate excretion is induced, cell growth slows and eventually stops, while sugars are mainly converted to glutamic acid. Another frequently observed metabolite that is excreted in significant concentrations is trehalose.

The kinetic model used to simulate the fermentation process is a physiological model that distinguishes between the two metabolic states of bacteria, in other words, the growth state and the metabolite production state. This type of model includes specific kinetic laws for each of these states. In addition, it includes a kinetic law that quantifies the rate of cell transition between the growth and production states. The model presented below corresponds to a strain of heat-inducible corynebacteria. In this case, glutamic acid excretion is triggered by raising the temperature of the medium from 33 to 39°C.

8.4.2.1. *Kinetic laws for the growth state*

When corynebacteria cells are in their growth state at 33°C, their specific growth rate μ follows a Monod-type law as a function of sugar concentration S:

$$\mu = \frac{\mu_{max}\,S}{K_S + S} \qquad [8.5]$$

In this growth state, their specific sugar consumption rate q_{Sg} includes a term that is proportional to the growth rate and a sugar consumption term for cell maintenance:

$$q_{Sg} = Y_{S/X}\,\mu + m_{Sg} \qquad [8.6]$$

Similarly, their specific oxygen consumption rate q_{O2g} includes a term that is proportional to the growth rate and an oxygen consumption term for cell maintenance:

$$q_{O2g} = Y_{O2/X}\,\mu + m_{O2g} \qquad [8.7]$$

Raising the temperature from 33°C to 39°C causes the cells to transition to the second production state. The volumetric rate of transition is proportional to the concentration X_g of cells in the growth state:

$$r_{gp}''' = k_{gp} X_g \qquad [8.8]$$

8.4.2.2. *Kinetic laws for the production state*

Once the cells have transited through the 39°C production state, growth stops and they convert sugar into glutamate and trehalose. Their specific rate of glutamate production q_{Glu} can be limited by the sugar concentration S and inhibited by the glutamate concentration Glu. It is expressed by the following law:

$$q_{Glu} = q_{Glu,max}\left(\frac{S}{K_{S,Glu} + S}\right)\left(1 - \frac{Glu}{Glu_m}\right) \qquad [8.9]$$

In this law, glutamate inhibition is linear, characterized by a maximum concentration of glutamate Glu_m that can be excreted.

The simultaneous production of trehalose by corynebacteria is the result of a metabolic spillover phenomenon. Trehalose is excreted when, at a high sugar concentration in the medium, the incoming sugar flux is too high to be converted into glutamate. The specific trehalose production rate q_{Tr} is expressed as a function of sugar concentration by a sigmoidal law that simulates the effect of sugar thresholds on trehalose production:

$$q_{Tr} = q_{Tr,max}\left(\frac{S^5}{K_{Tr} + S^5}\right) \qquad [8.10]$$

In this state of metabolite production, the specific sugar consumption rate q_{Sp} comprises a term that is proportional to the glutamate production rate, a term that is proportional to the trehalose production rate and a sugar consumption term for cell maintenance:

$$q_{Sp} = Y_{Y_{S/Glu}} q_{Glu} + Y_{S/Tr}\, q_{Tr} + m_{Sp} \qquad [8.11]$$

Similarly, the specific oxygen consumption rate q_{O2p} includes a term that is proportional to the glutamate production rate, a term that is proportional to the trehalose production rate, and an oxygen consumption term for cell maintenance:

$$q_{O2p} = Y_{O2/Glu}''\, q_{Glu} + Y_{O2/Tr} q_{Tr} + m_{O2p} \qquad [8.12]$$

For both physiological states, the kinetic parameters depend on the *Corynebacterium glutamicum* strain used. In the simulations used to optimize glutamate production, the kinetic parameters are shown in Table 8.4.

Growth stage		**Production state**	
μ_{max}: 0.55 h^{-1} K_S: 1 gS/L $Y_{S/X}$: 2.1 gS/gX m_{Sg}: 0.1 gS/X.h	$Y_{O2/X}$: 0.65 gO_2/gX m_{O2g}: 0.08 gO_2/X.h k_{gp}: 0.3 h^{-1}	$q_{Glu,max}$: 0.75 gGlu/gX.h K_{SGlu}: 0.5 gS/L Glu_m: 160 gGlu/L $q_{Tr,max}$: 0.13 gTr/gX.h K_{Tr}: 10^6 $(gS/L)^5$	$Y_{S/Glu}$: 1.4 gS/gX $Y_{S/Tr}$: 1.2 gS/gX m_{Sp}: 0.05 gS/X.h $Y_{O2/Glu}$: 0.4 gO_2/gX $Y_{O2/Tr}$: 0.35gO_2/gX m_{O2p}: 0.05 gO_2/X.h

Table 8.4. *Kinetic parameters of the glutamic acid fermentation simulation model*

8.4.3. *Optimum implementation of glutamic acid fermentation*

As previously explained in the chapter on aerobic fermentations, it is possible to produce a nongrowth-associated metabolite, such as glutamate, using a batch or fed-batch fermenter. However, glutamate production is not feasible in steady continuous mode due to the cell washout when growth is stopped.

In this section, we compare the optimum operating conditions for glutamic acid fermentation in batch and fed-batch modes. This comparison takes into account the use of a 200 m^3 volume aerated fermenter, with an aeration capacity characterized by an oxygen transfer coefficient $k_l a$ close to 400 h^{-1}. The aim of optimizing the fermentation process is to minimize the cost of glutamic acid production. From a technical point of view, economic optimization consists of finding the combination of operating variables that gives the highest values for fermenter productivity, sugar-to-glutamate conversion efficiency and final glutamate concentration, which influences the sizing of the downstream evaporation-concentration stage.

8.4.3.1. *Batch fermentation optimization*

For batch fermentation, the operational variables to be optimized are the initial sugar concentration (S_0), the induction time of glutamate excretion by temperature change (t_{ind}) and the final fermentation time (t_f). First, the induction time for the transition from the rapid growth state to the metabolite excretion state is optimized to achieve the highest cell concentration permitted by the oxygen transfer rate of the fermenter. This means that the peak oxygen consumption rate coincides with the maximum oxygen transfer rate. With this optimized induction time value, fermentation simulation combined with a glutamate production cost assessment indicates a minimum production cost for an initial sugar concentration of around 230 g/L and a fermentation time of around 33 h.

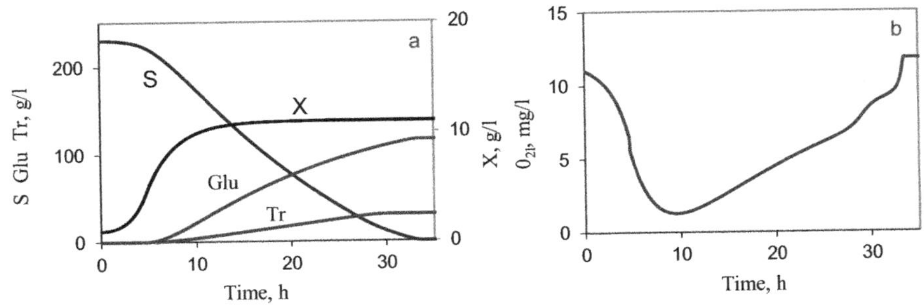

Figure 8.10. *(a) Variation in sugar S, biomass X, glutamate Glu and trehalose Tr concentrations during optimized batch fermentation. (b) Variation in dissolved oxygen O_{2l} during optimized batch fermentation*

Under these optimum conditions, variations in sugar, biomass, glutamate, trehalose and dissolved oxygen concentrations during fermentation are shown in Figures 8.10(a) and (b). The temperature change occurs at 4.5 h, allowing a maximum cell concentration of 11 g/L to be reached. Shortly after this temperature change, glutamate and trehalose excretion begins. Lastly, when all of the sugar has been consumed, glutamate concentration reaches 115 g/L, while trehalose is close to 30 g/L. Overall sugar conversion efficiency, fermenter productivity, annual glutamate and trehalose production, and glutamate production costs are shown in Table 8.5.

	X, g/L	Glu, g/L	$Y_{Glu/S\ glob}$, gGlu/gS	$Prod_{Glu}$, gGlu/L.h	Cost, €/kg	Annual Glu, tGlu	Annual Tr, tTr
Batch	11	115	0.5	2.8	2.70	4,300	1,000
Fed-batch	10.5	100	0.59	2.6	2.60	3,900	60

Table 8.5. *Technical and economic characteristics of optimized batch and fed-batch glutamate fermentation*

With the strain of *Corynebacterium glutamicum* used, the optimum cost of production in batch mode is around €2.7 per kg of glutamate. The main cost comes from raw materials, mainly sugar, which account for almost 50% of the production cost. Equipment contributes 20%, while energy and staff account for around 15% of the cost (Figure 8.11).

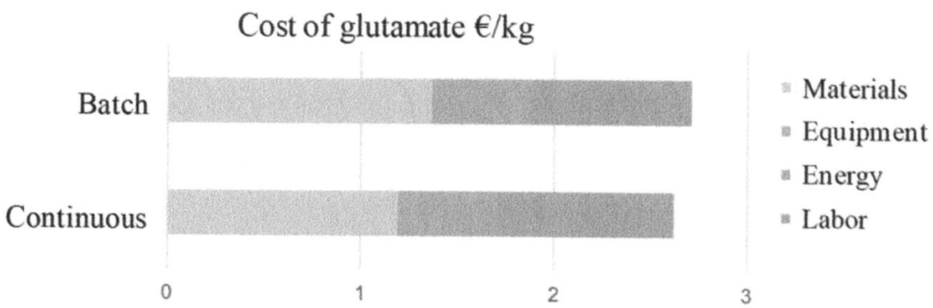

Figure 8.11. *Analysis of glutamic acid production costs in batch and fed-batch fermentation by origin. For a color version of this figure, see www.iste.co.uk/ghoul/bioreactor.zip*

8.4.3.2. *Optimizing fed-batch fermentation*

The main objective of fed-batch fermentation is to run the fermentation at a limiting sugar concentration in order to greatly reduce trehalose production compared to batch fermentation. Optimization of fed-batch fermentation involves six operational variables (Figure 8.12(a)): initial sugar concentration (S_0), initial fraction of medium volume (V_0/V_t), induction time (t_{ind}), feed sugar concentration (S_{feed}), feed time (t_{feed}) or steady sugar concentration (S_{ss}) and, lastly, final fermentation time (t_f). The methodology used to assess these variables is described in detail in the chapter on fermentations.

As in the case of batch fermentation, the induction time by temperature rise is initially determined in order to reach the maximum cell concentration allowed by oxygen transfer in the fermenter. The initial sugar concentration is also assessed according to the nutritional requirements of the cell to produce this maximum concentration. The feeding time is then set to reach the steady concentration, representing the critical threshold for the onset of trehalose excretion (Figure 8.12(b)). At this level of sugar concentration, the rate of glutamate production undergoes a slight reduction, while trehalose production is greatly reduced. The feed sugar concentration is determined as the highest available sugar concentration. The higher the S_{feed} value, the lower the volume of feed solution is added, the higher the initial volume of medium, and the greater the quantity of cells produced during the first growth phase. Ultimately, the initial fraction of medium in the fermenter and the duration of fermentation are optimized to achieve the lowest possible production cost.

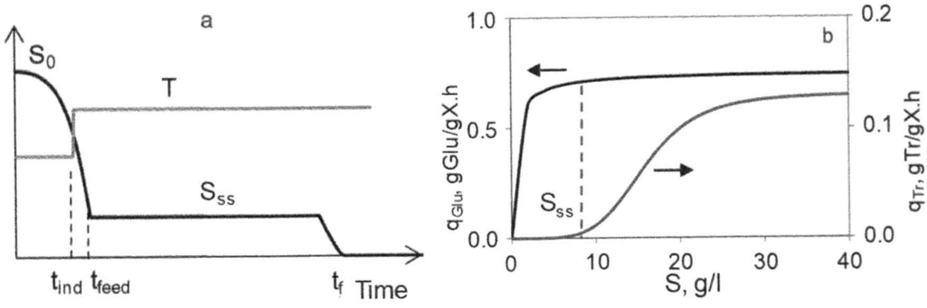

Figure 8.12. *(a) Operational variables of glutamate fed-batch fermentation. (b) Specific glutamate q_{Glu} and trehalose q_{Tr} production rates as a function of sugar S concentration. The dotted line indicates the optimum value for steady sugar S_{ss} concentration*

Under optimal operating conditions, the simulation of fed-batch fermentation is shown in Figures 8.13(a) and (b). Fermentation starts in batch mode with an initial media fraction of around 0.65, containing an initial sugar concentration of 40 g/L. The temperature change is carried out at 5.7 h, reaching a maximum cell concentration of around 10.5 g/L. The fermenter starts to be fed with a medium containing 400 g/L sugar is at 7.5 h, with the aim of stabilizing the sugar concentration at around 9 g/L. Feeding is stopped at 32 h, when the fermenter reaches its maximum medium capacity. At this point, glutamate concentration rises to 105 g/L, while trehalose concentration is 1.5 g/L.

For this fed-batch fermentation, the overall sugar conversion efficiency, fermenter productivity, annual metabolite production and glutamate production cost are shown in Table 8.5. Compared to batch fermentation, the main advantage of fed-batch is its capacity to drastically reduce trehalose production. Limiting sugar concentration, responsible for the reduction in trehalose production, also leads to a reduction in fermenter productivity of around 10%, resulting in lower annual glutamate production. However, fed-batch operation leads to a significant increase in sugar-to-glutamate conversion efficiency, from 0.50 to 0.59. This reduction in sugar consumption per kilogram of glutamate produced translates into a decrease in the cost of glutamate production, from 2.70 € to 2.60 € per kilogram. The main component of the production cost comes from raw materials, which account for over 45% of the total cost. Equipment and staff cost contribute around 20%, while energy accounts for around 15% (Figure 8.11).

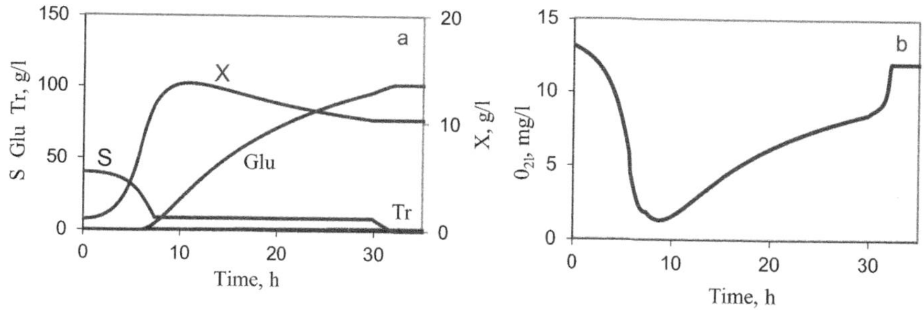

Figure 8.13. *(a) Variation in the concentrations of sugar S, biomass X, glutamate Glu, trehalose Tr and dissolved oxygen O_{2l} during optimized batch fermentation. (b) Variation in dissolved oxygen O_{2l} during optimized batch fermentation*

8.4.4. *Fermentation intensification strategy*

Glutamate fermentation productivity is limited by the biomass concentration that can be achieved during the initial growth phase. This biomass concentration, around 10 g/L, is the maximum allowed by the oxygen transfer rate in the fermenter. At the specified oxygen transfer coefficient of the fermenter, the maximum oxygen transfer rate depends on the average oxygen solubility in the fermenter. It can be increased by increasing the mole fraction of oxygen in the inlet air flow, and thus the partial pressure of oxygen in the fermenter.

Oxygen-enriched air technology consists of adding a gaseous phase containing a high molar fraction of oxygen to the inlet air, which can be obtained by passing the air through a bed of molecular sieves that absorb nitrogen molecules. The most economical method of operating the fermenter is to only add the oxygen molecules required at each moment of fermentation, so as to prevent oxygen limitations in the medium. This is achieved by installing an oxygen regulator on the fermenter, through which additional oxygen is added to prevent the oxygen level from falling below the minimum $O_{21,min}$ level. Under these conditions, the dissolved oxygen level is maintained at the $O_{21,min}$ value throughout the period of adding oxygen-enriched air, and the corresponding maximum oxygen transfer rate is automatically adjusted to the oxygen consumption rate in the fermenter (Figures 8.14(a) and (b)).

Simulation results for fed-batch fermentation with dissolved oxygen control are shown in Table 8.6. Oxygen enrichment leads to a sharp increase in biomass concentration, from 11 to 19 g/L. This translates into a 25% increase in fermenter productivity and annual glutamate production. At this biomass level, productivity becomes limited by the fermenter's heat removal capacity. Indeed, increasing the rate of oxygen consumption simultaneously increases the generation of metabolic heat in the fermenter. This is possible up to the maximum heat transfer rate from the agitated medium to the external coolant. From an economic point of view, the increase in annual production can be achieved without any significant change in the production cost per kilogram of glutamate, as the additional cost of the added oxygen is offset by the increase in annual glutamate production.

	X, g/L	Glu, g/L	$Y_{glu/S\ glob}$, gGlu/gS	$Prod_{Glu}$, gGlu/L.h	Cost, €/kg	Annual glu, tGlu	Annual Tr tTr
Fed-batch + oxygen	19	90	0.51	4.6	2.80	4,900	90

Table 8.6. *Technical and economic characteristics of glutamate fed-batch fermentation with oxygen enrichment*

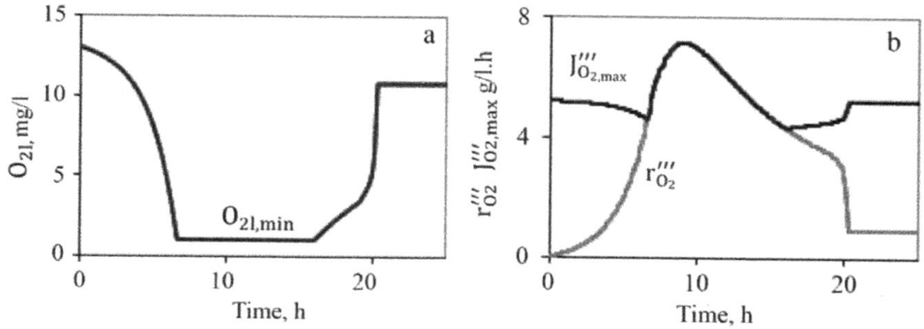

Figure 8.14. *(a) Variation in dissolved oxygen concentration O_{2l} during fed-batch fermentation with oxygen enrichment. (b) Variation in oxygen consumption volumetric rate r_{O2}''' and maximum oxygen transfer volumetric rate $J_{O2}'''{}_{max}$ during fed-batch fermentation with oxygen enrichment*

8.4.5. *Environmental optimization of the glutamic acid fermentation process*

Increases in fermenter productivity and sugar conversion efficiency, implemented to reduce the cost of glutamate production, also contribute to reducing the environmental impact indicators of the glutamate produced. Several additional water and heat recycling approaches can also be implemented for the entire production process in order to reduce the three main environmental impacts, namely, WAT, PEC and GWP (Figure 8.15).

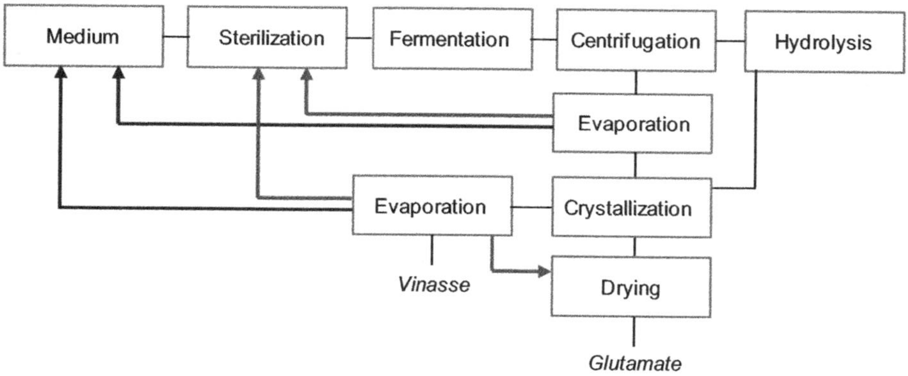

Figure 8.15. *Water and heat recycling strategies for the glutamic acid production process. For a color version of this figure, see www.iste.co.uk/ghoul/bioreactor.zip*

The glutamate production process is characterized by a high WAT indicator, due to the water used to prepare the fermentation media and the high WAT for the processed sugar. The WAT indicator for glutamate can be reduced by using the water produced during the two evaporation stages of glutamate solution and crystallization vinasses.

Sterilization, evaporation and drying operations require large amounts of heat. The PEC indicator can be reduced by using part of the steam and condensate phases from evaporation operations as heat inputs for sterilization and drying operations.

The GWP impact indicator for glutamate essentially comes from CO_2 emissions during fermentation and the use of gas for heating. While it is not feasible to recover the CO_2 present at low levels in the air leaving the fermenter, thermal integration between process operations, by reducing gas consumption, also helps to reduce the GWP indicator.

Assessment of the impact of these modifications on the three impact indicators of the glutamate production process shows a reduction of almost 15% in PEC, 40% in WAT and almost 20% in GWP. Under these optimized conditions, glutamate's environmental footprint is characterized by three impact indicator values per kilogram of glutamate produced: 30 MJ/kg for PEC, 13 kg water/kg for WAT and around 1 kg CO_2/kg for GWP.

8.5. References

Ariyajaroenwong, P., Laopaiboon, P., Salakkam, A., Srinophakun, P., Laopaiboon, L. (2016). Kinetic models for batch and continuous ethanol fermentation from sweet sorghum juice by yeast immobilized on sweet sorghum stalks. *Journal of the Taiwan Institute of Chemical Engineers*, 66, 210–216.

Bayrock, D.P. and Michael Ingledew, W. (2001). Application of multistage continuous fermentation for production of fuel alcohol by very-high-gravity fermentation technology. *Journal of Industrial Microbiology and Biotechnology*, 27(2), 87–93.

Cardona, C.A. and Sánchez, Ó.J. (2007). Fuel ethanol production: Process design trends and integration opportunities. *Bioresource Technology*, 98(12), 2415–2457.

Chauvel, A., Fournier, G., Raimbault, C. (2003). *Manual of Process Economic Evaluation.* Éditions Technip, Paris.

Delaunay, S., Gourdon, P., Lapujade, P., Mailly, E., Oriol, E., Engasser, J.M., Goergen, J.L. (1999). An improved temperature-triggered process for glutamate production with Corynebacterium glutamicum. *Enzyme and Microbial Technology*, 25(8–9), 762–768.

Geddes, C.C., Nieves, I.U., Ingram, L.O. (2011). Advances in ethanol production. *Current Opinion in Biotechnology*, 22(3), 312–319.

Hauschild, M.Z., Rosenbaum, R.K., Olsen, S.I. (2018). *Life Cycle Assessment*. Springer, Cham. doi: 10.1007/978-3-319-56475-3.

Klöpffer, W. and Grahl, B. (2014). *Life Cycle Assessment (LCA): A Guide to Best Practice*. John Wiley & Sons, New York.

Kumar, R., Vikramachakravarthi, D., Pal, P. (2014). Production and purification of glutamic acid: A critical review towards process intensification. *Chemical Engineering and Processing: Process Intensification*, 81, 59–71.

Peters, M.S., Timmerhaus, K.D., West, R.E. (2003). *Plant Design and Economics for Chemical Engineers*, 4th edition. McGraw-Hill, New York.

Puligundla, P., Smogrovicova, D., Mok, C., Obulam, V.S.R. (2019). A review of recent advances in high gravity ethanol fermentation. *Renewable Energy*, 133, 1366–1379.

Shaked, S., Crettaz, P., Saade-Sbeih, M., Jolliet, O., Jolliet, A. (2015). *Environmental Life Cycle Assessment*. CRC Press, Boca Raton.

Silla, H. (2003). *Chemical Process Engineering: Design and Economics*. CRC Press, Boca Raton.

Towler, G. and Sinnott, R. (2021). *Chemical Engineering Design: Principles, Practice and Economics of Plant and Process Design*. Butterworth-Heinemann, Oxford.

Wendisch, V.F., Jorge, J.M., Pérez-García, F., Sgobba, E. (2016). Updates on industrial production of amino acids using Corynebacterium glutamicum. *World Journal of Microbiology and Biotechnology*, 32, 1–10.

Zhang, X.W., Sun, T., Sun, Z.Y., Liu, X., Gu, D.X. (1998). Time-dependent kinetic models for glutamic acid fermentation. *Enzyme and Microbial Technology*, 22(3), 205–209.

Conclusion

Mohamed GHOUL
Laboratoire Réaction et Génie des Procédés (LRGP),
CNRS, Université de Lorraine, Nancy, France

The optimization of bioprocesses using bioreactors is a delicate task, requiring compromises between several factors governing bioreactor performance. This optimization must also consider upstream and downstream operations of the bioreactors, as well as the environmental impact of these processes.

The key factors to be considered in these processes are related to the choice of medium composition, bioreactor technology and operating procedure, control of the kinetics of the microorganisms of interest, and heat and matter transfer.

The choice and optimization of medium composition (carbon source, nitrogen source, phosphorus source, growth factors, source and luminosity in the case of microalgae) are crucial, as they not only influence the performance of these processes in terms of metabolite concentration, efficiency and productivity, but they also influence their environmental impact and production costs.

When it comes to bioreactor technology, the bioprocess engineer has to choose wisely from numerous implementation options (batch, continuous, semi-continuous, etc.). These choices must consider the kinetics of the microorganisms used and the limitations due to the transfer of matter, particularly oxygen.

To succeed in these various tasks, the bioprocess engineer must use bioreactor modeling-simulation approaches. This approach makes it possible to predict and compare the course of a process under various conditions, thus reducing the need for costly experiments. The most relevant models are those that integrate material and heat balances, as well as thermodynamic and kinetic laws to describe the fundamental processes governing bioreactor performance.

It is also important for the bioprocess engineer to bear in mind that there is no universal solution for bioprocess design. Each case is unique, and technological choices must be specifically adapted to the bioprocessing reaction, cell characteristics and industrial context.

The aim of this book is to provide researchers, students and industrialists with a set of bioreactor engineering tools for the development of optimal, competitive and environmentally friendly bioprocesses.

List of Authors

Jihène AMMAR
Laboratoire de Biotechnologie Bleu
et Bioproduits Aquatiques (B3 Aqua)
Institut National des Sciences et
Technologies de la Mer (INSTM)
Monastir Annex
Tunisia

Hatem BEN OUADA
Laboratoire de Biotechnologie Bleu
et Bioproduits Aquatiques (B3 Aqua)
Institut National des Sciences et
Technologies de la Mer (INSTM)
Monastir Annex
Tunisia

Jean-Marc ENGASSER
Laboratoire Réaction et Génie des
Procédés (LRGP)
CNRS
Université de Lorraine
Nancy
France

Mohamed GHOUL
Laboratoire Réaction et Génie des
Procédés (LRGP)
CNRS
Université de Lorraine
Nancy
France

Franck JOLIBERT
Musée Français de la Brasserie
Saint-Nicolas-de-Port
France

Céline LOUBIÈRE
Laboratoire Réaction et Génie des
Procédés (LRGP)
CNRS
Université de Lorraine
Nancy
France

Eric OLMOS
Laboratoire Réaction et Génie des
Procédés (LRGP)
CNRS
Université de Lorraine
Nancy
France

Index

F

G, H, I

K, L

M

O

P

R

S

T, Y

Printed and bound by CPI Group (UK) Ltd, Croydon, CR0 4YY
13/09/2026
14961069-0004